Time Delay ODE/PDE Models

Models

Applications in Biomedical Science and Engineering

Time Delay ODE/PDE Models

Applications in Biomedical Science and Engineering

W. E. Schiesser

CRC Press
Taylor & Francis Group
Boca Raton London New York

CRC Press is an imprint of the
Taylor & Francis Group, an **informa** business

CRC Press
Taylor & Francis Group
6000 Broken Sound Parkway NW, Suite 300
Boca Raton, FL 33487-2742

First issued in paperback 2023

ISBN-13: 978-0-367-42797-9 (hbk)
ISBN-13: 978-1-03-265432-4 (pbk)
ISBN-13: 978-0-367-42798-6 (ebk)

DOI: 10.1201/9780367427986

Publisher's Note
The publisher has gone to great lengths to ensure the quality of this reprint but points out that some imperfections in the original copies may be apparent.

Library of Congress Cataloging-in-Publication Data

Names: Schiesser, W. E., author.
Title: Time delay ODE/PDE models : applications in biomedical science and engineering / by William E. Schiesser.
Description: Boca Raton : CRC Press, [2020] | Includes bibliographical references and index. | Summary: "The intent of this book is to present a methodology for the formulation and computer implementation of mathematical models based on time delay ordinary differential equations (DODEs) and partial differential equations (DPDEs) in biomedical science and engineering (BMSE)"– Provided by publisher.
Identifiers: LCCN 2019039302 (print) | LCCN 2019039303 (ebook) | ISBN 9780367427979 (hardback ; alk. paper) | ISBN 9780367427986 (ebook)
Subjects: MESH: Numerical Analysis, Computer-Assisted | Models, Theoretical | Mathematical Computing | Biomedical Engineering
Classification: LCC RC262 (print) | LCC RC262 (ebook) | NLM QT 35 | DDC 616.99/400285–dc23
LC record available at https://lccn.loc.gov/2019039302
LC ebook record available at https://lccn.loc.gov/2019039303

Visit the Taylor & Francis Web site at
http://www.taylorandfrancis.com

and the CRC Press Web site at
http://www.crcpress.com

Contents

Preface

Time delayed (lagged) variables are an inherent feature of biological/physiological systems. For example, infection from a disease may at first be asymptomatic, and only after a delay is the infection apparent so that treatment can begin. Thus, to adequately describe physiological systems, time delays are frequently required and must be included in the equations of mathematical models. That is, the equations cannot include just concurrent terms representing phenomena/mechanisms at a particular time.

The intent of this book is to present a methodology for the formulation and computer implementation of mathematical models based on time delay ordinary differential equations (DODEs) and partial differential equations (DPDEs). The DODE/DPDE methodology is presented through a series of example applications, particularly in biomedical science and engineering (BMSE). The computer-based implementation of the example models is presented through routines coded (programmed) in R, a quality, open-source scientific computing system that is readily available from the Internet. Formal mathematics is minimized, for example, no theorems and proofs. Rather, the presentation is through detailed examples that the reader, researcher, and analyst can execute on modest computers.

Chapter 1 has four example DODE applications that illustrate various properties of delayed systems. This discussion is then extended in Chapter 2 to an introductory DPDE model with the model solution presented as spatiotemporal numerical and 2D and 3D graphical (plotted) outputs. Subsequent chapters pertain to a series of DPDE BMSE applications.

In summary, the book is intended as an introduction to DODE/DPDE models through the use of R routines that are available as a download, without having to first study numerical methods (algorithms) and programming. The DPDE analysis is based on the method of lines (MOL), an established general algorithm for PDEs, implemented with finite differences. The example applications can first be executed to confirm the reported solutions, then extended by variation of the parameters and the equation terms, and even the formulation and use of alternative DODE/DPDE models.

The author would welcome comments/suggestions concerning this approach to time delayed systems (directed to: wes1@lehigh.edu).

W. E. Schiesser
Bethlehem, Pennsylvania

Author

W. E. Schiesser, PhD, is Emeritus McCann Professor in the Chemical and Biomolecular Engineering Department as well as a former professor in the Mathematics Department at Lehigh University in Bethlehem, Pennsylvania. He holds a PhD from Princeton University, and a honorary ScD from the University of Mons, Belgium. He is the author or co-author of a series of books in his field of research on numerical methods and associated software for ordinary, differential-algebraic, and partial differential equations (ODE/DAE/PDEs) and the development of mathematical models based on ODE/DAE/PDEs. More recently, Dr. Schiesser has authored several books on computer-based solutions to models for real-life phenomena, such as the development of Parkinson's disease.

1

Introduction to Delay Ordinary Differential Equations

Introduction

Example delay ordinary differential equations (DODEs) are considered in this chapter to explain how DODEs can be integrated numerically within the basic R system.[1] The R routines are discussed in detail and the solutions are displayed numerically and graphically. Generally, the routines include (1) a main program that calls a R library DODE integrator and (2) a subordinate routine that defines the DODE system.

1.1   Example 1: A basic DODE

The following classical ODE without delay provides a starting point for the subsequent discussion of DODEs.

$$\frac{dy}{dt} + \lambda y = 0; \; y(t_0) = y_0$$

The initial value of t, t_0, and the eigenvalue λ are specified constants.
   The solution for $t_0 = 0$

$$y(t) = y_0 e^{-\lambda t}$$

is smooth for derivatives of all orders, which contrasts with DODE solutions as illustrated by the following example [2, p. 124].

$$\frac{dy(t)}{dt} + y(t-1) = 0; \quad y(t) = 1, \text{ for } t \in [-1, 0] \qquad (1.1\text{-}1,2)$$

The derivative $dy(t)/dt$ is now dependent on past values of the dependent variable, $y(t-1)$, which in turn requires an initial condition (IC) that is specified over the interval $-1 \le t \le 0$. The IC is usually taken as y_0 (but this

---

[1]R is an open-source, quality scientific computing system that is readily available from the Internet [2].

particular *history* is not necessarily required and other functions of t can be used).

The solution to eq. (1.1-1) is[2] [2, p. 124]

$$
y(t) = \begin{cases}
1 - t, & 0 \le t \le 1 \\
t^2/2 - 2t + 3/2, & 1 \le t \le 2 \\
-t^3/6 + (3/2)t^2 - 4t + 17/6, & 2 \le t \le 3
\end{cases} \qquad (1.1\text{-}3)
$$

which is compared to the numerical solution in the main program discussed subsequently. The piecewise polynomial solution of eq. (1.1-3) contrasts with the exponential solution of the ODE with no delay, $y(t) = y_0 e^{-\lambda t}$ (from the beginning of Section 1.1). The differences in these solutions will be clear when the numerical and graphical output for eqs. (1.1-1,2) are discussed subsequently.

Eq. (1.1-3) can be verified as a solution to eqs. (1.1-1,2) by substitution. For example, for $t \in [1, 2]$,

$$
\begin{array}{ccc}
\text{eq. (1.1-1)} & \text{from solution} \\[6pt]
\dfrac{dy(t)}{dt} & t - 2 \\[10pt]
+y(t - 1) & \begin{array}{l} 1 - (t - 1) \\ = -(t - 2) \end{array} \\[12pt]
\text{sum} & \text{sum} \\
0 & 0
\end{array}
$$

1.1.1   Main program

A main program for eqs. (1.1-1,2) follows.

```
#
# Basic DODE
#
# Delete previous workspaces
  rm(list=ls(all=TRUE))
#
# Access DODE integrator
  library(deSolve)
#
# Access functions for numerical solution
  setwd("f:/dpde/chap1/ex1");
```

[2]The contribution of the solutions to eqs. (1.1-1,2) by Professor Yue Yu, Department of Mathematics, Lehigh University, is gratefully acknowledged.

```
  source("ode1a.R");
#
# Parameters
  tau=1;
#
# Temporal grid
  nout=101;t0=0;tf=10;
  times=seq(from=t0,to=tf,by=(tf-t0)/(nout-1));
#
# IC vector
  y0=rep(0,1);
  y0[1]=1;
  ncall=0;
#
# Integration of DODE
  yout=dede(y=y0,times=times,func=ode1a);
  nrow(yout);
  ncol(yout);
#
# Vectors for DODE solution
  yp=rep(0,nout);
  tp=rep(0,nout);
  for(it in 1:nout){
    tp[it]=yout[it,1];
    yp[it]=yout[it,2];
  }
#
# Display detailed numerical solution
# print(yout);
#
# Display numerical solution, and for tau=1, analytical
# solution, difference
  if(tau==1){
#
#   Include analytical solution for 0 <= t <= 3
    ya=rep(0,31);diff=rep(0,31);
    for(it in 1:31){
      if(it <= 11){ya[it]=
        1-tp[it];}
      if((it >  11)&(it <= 21)){ya[it]=
        tp[it]^2/2-2*tp[it]+3/2;}
      if((it >  21)&(it <= 31)){ya[it]=
        -tp[it]^3/6+(3/2)*tp[it]^2-4*tp[it]+17/6;}
    }
    cat(sprintf(
      "\n   it      t       y(t)      ya(t)     diff(t)\n"));
    for(it in 1:31){
      diff[it]=yp[it]-ya[it];
      cat(sprintf("\n %4d %6.2f %8.3f %8.3f %10.3e",
```

```
                       it,tp[it],yp[it],ya[it],diff[it]));}
    }
#
#   For t > 3, only numerical solution
    for(it in 32:101){
       cat(sprintf("\n %4d %6.2f %8.3f ",
                    it,tp[it],yp[it]));
  }
#
# Display calls to ode1a
  cat(sprintf("\n ncall = %4d\n",ncall));
#
# Plot DODE solution against t
  par(mfrow=c(1,1));
  plot(tp,yp,type="l",xlab="t",ylab="y(t)",
       lty=1,main="",lwd=2,col="black");
```

Listing 1.1   Main program for eqs. (1.1-1,2).

We can note the following details about Listing 1.1.

- Previous workspaces are deleted.

```
    #
    # Basic DODE
    #
    # Delete previous workspaces
      rm(list=ls(all=TRUE))
```

- The R ODE integrator library deSolve is accessed. Then the directory with the files for the solution of eqs. (1.1-1,2) is designated. Note that setwd uses / rather than the usual \.

```
    #
    # Access DODE integrator
      library(deSolve)
    #
    # Access functions for numerical solution
      setwd("f:/dpde/chap1/ex1");
      source("ode1a.R");
```

ode1a.R is the routine for eqs. (1.1-1,2) (discussed subsequently).

- The delay in eq. (1.1-1) is defined.

```
    #
    # Parameters
      tau=1;
```

- An interval in t of 101 points is defined for $0 \le t \le 10$ so that tout=0,0.1,...,10.

```
#
# Temporal grid
  nout=101;t0=0;tf=10;
  times=seq(from=t0,to=tf,by=(tf-t0)/(nout-1));
```

- The history (IC) for eq. (1.1-1) is specified as eq. (1.1-2).

```
#
# IC vector
  y0=rep(0,1);
  y0[1]=1;
  ncall=0;
```

Also, the counter for the number of calls to ode1a is initialized.

- Eq. (1.1-1) is integrated (solved numerically) by the library integrator dede (available in deSolve, [2, Chapter 7]). As expected, the inputs to dede are the IC vector y0, the vector of output values of t, times, and the DODE function, ode1a. The length of u0 (1) informs dede how many DODEs are to be integrated. y, times, func are reserved names.

```
#
# Integration of DODE
  yout=dede(y=y0,times=times,func=ode1a);
  nrow(yout);
  ncol(yout);
```

The numerical solution to eq. (1.1-1) is returned in matrix yout. In this case, yout has the dimensions *nout* $\times (1+1) = 101 \times 1 + 1 = 2$, which are confirmed by the output from nrow(out),ncol(out) (included in the numerical output considered subsequently).

The offset $1 + 1$ is required since the first element of each of the 101 rows has the output t (also in times), and the second element has the solution of eq. (1.1-1).

- Vectors are defined for the computed eq. (1.1-1) solution (in array yout returned by dede). The solution is then placed in these arrays.

```
#
# Vectors for DODE solution
  yp=rep(0,nout);
  tp=rep(0,nout);
  for(it in 1:nout){
    tp[it]=yout[it,1];
    yp[it]=yout[it,2];
  }
```

Again, ,2 is required in yp[it]=yout[it,2] since the first element of each solution vector (for a particular index it) is the value of t associated with the solution.

- The complete solution in `yout` can be displayed with `print` (but use with caution since this can display many numbers).

```
#
# Display detailed numerical solution
# print(yout);
```

- For $\tau = 1$, the numerical and analytical eq. (1.1-3), and difference, are displayed for $0 \le t \le 3$.

```
#
# Display numerical solution, and for tau=1, analytical
# solution, difference
  if(tau==1){
#
#     Include analytical solution for 0 <= t <= 3
      ya=rep(0,31);diff=rep(0,31);
      for(it in 1:31){
        if(it <= 11){ya[it]=
          1-tp[it];}
        if((it >  11)&(it <= 21)){ya[it]=
          tp[it]^2/2-2*tp[it]+3/2;}
        if((it >  21)&(it <= 31)){ya[it]=
          -tp[it]^3/6+(3/2)*tp[it]^2-4*tp[it]+17/6;}
      }
      cat(sprintf(
        "\n   it       t      y(t)     ya(t)      diff(t)\n"));
      for(it in 1:31){
        diff[it]=yp[it]-ya[it];
        cat(sprintf("\n %4d %6.2f %8.3f %8.3f %10.3e",
                    it,tp[it],yp[it],ya[it],diff[it]));}
      }
```

- For $3 < t \le 10$, only the numerical solution is displayed (the analytical solution of eq. (1.1-3) does not apply to this interval in t).

```
#
#     For t > 3, only numerical solution
      for(it in 32:101){
        cat(sprintf("\n %4d %6.2f %8.3f ",
                    it,tp[it],yp[it]));
    }
```

- The number of calls to `ode1a` is displayed at the end of the solution.

```
#
# Display calls to ode1a
  cat(sprintf("\n ncall = %4d\n",ncall));
```

- The numerical solution to eq. (1.1-1) is plotted against t.

```
#
# Plot DODE solution against t
  par(mfrow=c(1,1));
  plot(tp,yp,type="l",xlab="t",ylab="y(t)",
       lty=1,main="",lwd=2,col="black");
```

This completes the discussion of the main program of Listing 1.1. Routine ode1a called dede in the main program is considered next.

1.1.2   DODE routine

ode1a follows.

```
  ode1a=function(t,y,parm,lag) {
#
# DODE source term
  if (t > tau){
    ylag=lagvalue(t-tau);
  } else {
    ylag=y0;
  }
#
# DODE
  yt=rep(0,1);
  yt[1]=-ylag[1];
#
# Increment calls to ode1a
  ncall <<- ncall+1;
#
# Return DODE t derivative vector
  return(list(c(yt)));
#
# End of ode1a
  }
```

Listing 1.2   DODE routine for eqs. (1.1-1,2).

We can note the following details about Listing 1.2.

- The function is defined.

  ```
  ode1a=function(t,y,parm,lag) {
  ```

  t is the current value of t in eqs. (1.1-1,2). y is the current numerical solution to eq. (1.1-1). parm is an argument to pass parameters to ode1a (unused, but required in the argument list). lag is the lag (delay) of an DODE system, which is unused in the current application (the lag is passed to ode1a from the main program of Listing 1.1 as parameter tau). The arguments must be listed in the order stated to properly interface with dede called in the main program of Listing 1.1. The derivative

$dy(t)/dt$ of eq. (1.1-1) is calculated and returned to `dede` as explained subsequently.

- The delay term $y(t-1)$ in eq. (1.1-1) is programmed.

```
#
# DODE source term
  if (t > tau){
    ylag=lagvalue(t-tau);
  } else {
    ylag=y0;
  }
```

The `if` selects either of two cases:

- `if (t > tau)`: The current value of t (from the first input argument of `ode1a`) is greater than the delay `tau`. The lag variable $y(t-1)$ is placed in `lag`.

- `} else {` : The current value of t is less than or equal to `tau` so that history in `y0` (from the main program of Listing 1.1) is used.

`lagvalue` is an R utility that defines past values of `y` (the second input argument of `ode1a`). `ylag` is a vector with the past values of `y` that can be used in the following programming of eq. (1.1-1).

- Eq. (1.1-1) is programmed. The derivative $dy(t)/dt$ is placed in `yt[1]`.

```
#
# DODE
  yt=rep(0,1);
  yt[1]=-ylag[1];
```

- The counter for the calls to `ode1a` is incremented and returned to the main program of Listing 1.1 by «-.

```
#
# Increment calls to ode1a
  ncall <<- ncall+1;
```

- The derivative `yt` is returned to `dede` for the next step along the solution.

```
#
# Return DODE t derivative vector
  return(list(c(yt)));
```

The derivative `yt` is returned as a `list` as required by `dede`. `c` is the R vector utility.

- The final } concludes ode1a.

```
#
# End of ode1a
  }
```

The numerical and graphical (plotted) output from the main program of Listing 1.1 and DODE routine of Listing 1.2 is considered next.

1.1.3 Numerical, graphical output

Abbreviated numerical output is given in Table 1.1.

```
[1] 101

[1] 2

it      t       y(t)     ya(t)     diff(t)
 1    0.00    1.000    1.000   0.000e+00
 2    0.10    0.900    0.900   1.110e-16
 3    0.20    0.800    0.800   0.000e+00
 4    0.30    0.700    0.700   1.110e-16
 5    0.40    0.600    0.600   1.110e-16
 6    0.50    0.500    0.500   1.110e-16
 7    0.60    0.400    0.400   1.110e-16
 8    0.70    0.300    0.300   1.665e-16
 9    0.80    0.200    0.200   2.776e-16
10    0.90    0.100    0.100   3.747e-16
11    1.00    0.000    0.000   4.580e-16
                         .        .
                         .        .
                         .        .
       Output for t=1.1 to 1.9 removed
                         .        .
                         .        .
                         .        .
21    2.00   -0.500   -0.500   1.389e-06
22    2.10   -0.495   -0.495  -3.317e-06
23    2.20   -0.481   -0.481  -3.456e-06
24    2.30   -0.460   -0.460  -3.595e-06
25    2.40   -0.431   -0.431  -3.734e-06
26    2.50   -0.396   -0.396  -3.873e-06
27    2.60   -0.356   -0.356  -4.011e-06
28    2.70   -0.312   -0.312  -4.150e-06
29    2.80   -0.265   -0.265  -4.289e-06
30    2.90   -0.217   -0.217  -4.428e-06
31    3.00   -0.167   -0.167  -4.567e-06
32    3.10   -0.117
33    3.20   -0.068
```

```
34    3.30    -0.021
35    3.40     0.024
36    3.50     0.065
37    3.60     0.103
38    3.70     0.136
39    3.80     0.165
40    3.90     0.189
41    4.00     0.208
        .        .
        .        .
        .        .

Output for t=4.1 to 8.9
        removed

        .        .
        .        .
        .        .
91    9.00     0.053
92    9.10     0.053
93    9.20     0.052
94    9.30     0.050
95    9.40     0.047
96    9.50     0.044
97    9.60     0.040
98    9.70     0.035
99    9.80     0.031
100   9.90     0.025
101  10.00     0.020

ncall = 1163
```

Table 1.1   Abbreviated output from Listings 1.1 and 1.2

We can note the following details about this output.

- 101 t output points as the first dimension of the solution matrix yout from dede as programmed in the main program of Listing 1.1.

- The solution matrix yout returned by dede has two elements as a second dimension. The first element is the value of t. The second is $y(t)$ from eq. (1.1-1) (for each of the 101 output points).

- For $0 \le t \le 3$, the numerical and analytical solutions, and the difference, are displayed. In general, the agreement between the two solutions is within -4.567e-06 or better (less). This agreement confirms the numerical solution from dede.

- For $3 < t \le 10$, the numerical solution is displayed.

- The computational effort is modest, `ncall` = 1163, so that `dede` computed a solution to eqs. (1.1-1,2) efficiently and accurately.

The graphical output is in Figure 1.1-1.

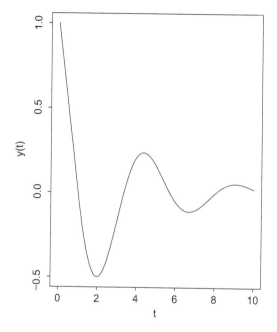

Figure 1.1-1   Numerical solution $y(t)$ from eqs. (1.1-1,2).

Figure 1.1-1 indicates the solution $y(t)$ is oscillatory over $0 \le t \le 10$ and approaches zero. This contrasts with the exponential solution of the ODE with no delay, $y(t) = y_0 e^{-\lambda t}$.

In summary, this example demonstrates the use of `dede` for the numerical integration of a DODE, and the characteristics of the solution that are different than for an ODE that has only concurrent terms in t. The next example illustrates some additional features in the use of `dede`.

1.2   Example 2: One DODE, effect of delay

The following DODE models the chaotic production of white blood cells (WBC) [2, pp. 125-126]

$$\frac{dy}{dt} = ay(t - \tau)\frac{1}{1 + y(t - \tau)^c} - by; \quad y(t) = 0.5 \text{ for } t \le 0 \qquad (1.2\text{-}1,2)$$

where

- $y(t)$: Current WBC density.

- t: Time.

- $y(t) = 0.5$ for $t \leq 0$: History of $y(t)$.

- $y(t - \tau)$: WBC density at τ time units in the past.

- τ: Time delay of WBC production.

- $ay(t - \tau)\dfrac{1}{1 + y(t - \tau)^c}$: Rate of introduction of WBC into the blood in response to demand at the previous time $t - \tau$.

- $-by$: Death rate of WBC.

- a, b, c: Constants to be specified.

- $\dfrac{dy}{dt}$: Rate of change of WBC density.

Of particular interest is the variation of the solutions of eqs. (1.2) with τ. This is considered for two cases in Table 1.2.

Case	Parameter values
case 1 deterministic	$a = 0.2; b = 0.1, c = 10, \tau = 10$
case 2 chaotic	$a = 0.2; b = 0.1, c = 10, \tau = 20$

Table 1.2   Parameters for eqs. (1.2)

The main program and DODE routines for eqs. (1.2) follows.

1.2.1   Main program

The following main program is similar to the main program in Listing 1.1, but some of the discussion is repeated here so that this example (eqs. (1.2)) is self-contained.

```
#
# White blood cells
#
# Delete previous workspaces
  rm(list=ls(all=TRUE))
#
# Access DODE integrator
  library(deSolve)
```

```
#
# Access functions for numerical solution
  setwd("f:/dpde/chap1/ex2");
  source("ode1a.R");
#
# Parameters
  a=0.2;
  b=0.1;
#
# Temporal grid
  nout=3001;t0=0;tf=300;
  times=seq(from=t0,to=tf,by=(tf-t0)/(nout-1));
#
# IC vector
  y0=rep(0,1);
  y0[1]=0.5;
  ncall=0;
#
# Integration of DODE
  yout1=dede(y=y0,times=times,func=ode1a,tau=10)
  nrow(yout1);
  ncol(yout1);
  yout2=dede(y=y0,times=times,func=ode1a,tau=20)
  nrow(yout2);
  ncol(yout2);
#
# Display calls to ode1a
  cat(sprintf("\n ncall = %4d\n",ncall));
#
# Plot DODE for tau=10,20
  plot(yout1,lwd=2,main="tau=10",
       ylab="y",mfrow=c(2, 2),which=1);
  plot(yout1[,-1],type="l",lwd=2,xlab="y");
  plot(yout2, lwd=2,main="tau=20",
       ylab="y",mfrow=NULL,which=1);
  plot(yout2[,-1],type="l",lwd=2,xlab="y");
```

Listing 1.3 Main program for eqs. (1.2).

We can note the following details about Listing 1.3.

- Previous workspaces are deleted.

```
#
# White blood cells
#
# Delete previous workspaces
  rm(list=ls(all=TRUE))
```

- The R ODE integrator library `deSolve` is accessed. Then the directory with the files for the solution of eqs. (1.2) is designated. Note that `setwd` uses / rather than the usual \.

```
#
# Access DODE integrator
  library(deSolve)
#
# Access functions for numerical solution
  setwd("f:/dpde/chap1/ex2");
  source("ode1a.R");
```

  `ode1a.R` is the routine for eqs. (1.2-1,2) (discussed subsequently).

- The parameters in eqs. (1.2) are defined.

```
#
# Parameters
  a=0.2;
  b=0.1;
```

- An interval in t of 3001 points is defined for $0 \leq t \leq 300$ so that `tout=0,0.1,...,300`.

```
#
# Temporal grid
  nout=3001;t0=0;tf=300;
  times=seq(from=t0,to=tf,by=(tf-t0)/(nout-1));
```

- The history (IC) for eq. (1.2-1) is specified as eq. (1.2-2).

```
#
# IC vector
  y0=rep(0,1);
  y0[1]=0.5;
  ncall=0;
```

  Also, the counter for the number of calls to `ode1a` is initialized.

- Eq. (1.2-1) is integrated by the library integrator `dede` (available in `deSolve`, [2, Chapter 7]). As expected, the inputs to `dede` are the IC vector `y0`, the vector of output values of t, `times`, the DODE function, `ode1a`, and the value of the delay in eq. (1.2-1), `tau`. The length of `u0` (1) informs `dede` how many DODEs are to be integrated. `y,times,func,tau` are reserved names so the four arguments can be placed in any order in the call to `dede`.

```
#
# Integration of DODE
  yout1=dede(y=y0,times=times,func=ode1a,tau=10)
```

```
        nrow(yout1);
        ncol(yout1);
        yout2=dede(y=y0,times=times,func=odela,tau=20)
        nrow(yout2);
        ncol(yout2);
```

Two solutions are computed with `tau=10,20`. The numerical solution to eq. (1.2-1) is returned in matrices `yout1,yout2`. In this case, `yout1,yout2` have the dimensions *nout* × (1 + 1) = 3001 × 1 + 1 = 2, which are confirmed by the output from `nrow(yout1),ncol(yout1)` and `nrow(yout2),ncol(yout2)` (included in the numerical output considered subsequently).

The offset 1 + 1 is required since the first element of each of the 3001 rows has the output *t* (also in `times`), and the second element has the solution of eq. (1.2-1).

- The number of calls to `odela` is displayed at the end of the solution.

```
#
# Display calls to odela
  cat(sprintf("\n ncall = %4d\n",ncall));
```

- The numerical solution to eq. (1.2-1) is plotted against *t*.

```
#
# Plot DODE for tau=10,20
  plot(yout1,lwd=2,main="tau=10",
       ylab="y",mfrow=c(2, 2),which=1);
  plot(yout1[,-1],type="l",lwd=2,xlab="y");
  plot(yout2, lwd=2,main="tau=20",
       ylab="y",mfrow=NULL,which=1);
  plot(yout2[,-1],type="l",lwd=2,xlab="y");
```

A 2 × 2 matrix of four plots is specified with `mfrow=c(2, 2)`. `which=1` specifies the plotting of *y*(*t*). In each case (`tau=10,20`), a second call to `plot` produces a plot of the delay variable *y*(*t* − *τ*) against *t* with `yout1[,-1],yout2[,-1]` and `[,` specifying all values of *t*.

This completes the discussion of the main program of Listing 1.3. The DODE routine `odela` is considered next.

1.2.2 DODE routine

`odela` follows.

```
  odela=function(t,y,parm,tau) {
#
# DODE source term
  if (t > tau){
```

```
    ylag=lagvalue(t-tau);
  } else {
    ylag=y0[1];
  }
#
# DODE
  yt=rep(0,1);
  yt[1]=a*ylag[1]*1/(1+ylag[1]^10)-b*y[1];
#
# Increment calls to ode1a
  ncall <<- ncall+1;
#
# Return DODE t derivative, lag variable
  return(list(yt[1],ylag[1]));
#
# End of ode1a
  }
```

Listing 1.4   DPDE routine for eqs. (1.2).

We can note the following details about Listing 1.4.

- The function is defined.

  ```
  ode1a=function(t,y,parm,tau) {
  ```

  t is the current value of t in eqs. (1.2). y is the current numerical solution to eq. (1.2-1). parm is an argument to pass parameters to ode1a (unused, but required in the argument list). tau is the lag (delay) of eq. (1.2-1). The arguments must be listed in the order stated to properly interface with dede called in the main program of Listing 1.3. The derivative $dy(t)/dt$ of eq. (1.2-1) is calculated and returned to dede as explained subsequently.

- The delay term $y(t - \tau)$ in eq. (1.2-1) is programmed.

  ```
  #
  # DODE source term
    if (t > tau){
      ylag=lagvalue(t-tau);
    } else {
      ylag=y0[1];
    }
  ```

  The if selects either of two cases:

  - if (t > tau): The current value of t (from the first input argument of ode1a) is greater than the delay tau. The lag variable $y(t - \tau)$ is placed in lag.
  - } else { : The current value of t is less than or equal to tau so that history in y0 (from the main program of Listing 1.3) is used.

lagvalue is a R utility that defines past values of y (the second input argument of ode1a). ylag is a vector with the past values of y that can be used in the following programming of eq. (1.2-1).

- Eq. (1.2-1) is programmed. The derivative $dy(t)/dt$ is placed in yt[1].

```
#
# DODE
  yt=rep(0,1);
  yt[1]=a*ylag[1]*1/(1+ylag[1]^10)-b*y[1];
```

- The counter for the calls to ode1a is incremented and returned to the main program of Listing 1.3 by «-.

```
#
# Increment calls to ode1a
  ncall <<- ncall+1;
```

- The derivative yt and the lag variable ylag are returned to dede for the next step along the solution.

```
#
# Return DODE t derivative, lag variable
  return(list(yt[1],ylag[1]));
```

yt and ylag are returned as a list as required by dede. c is not used (as in Listing 1.2) since two elements, yt[1],ylag[1], are being returned, but only one IC is specified in the main program of Listing 1.3 (the use of return(list(c(yt[1],ylag[1]))) generates an error message explaining the problem).

- The final } concludes ode1a.

```
#
# End of ode1a
  }
```

The numerical (Table 1.3) and graphical (plotted) output from the main program of Listing 1.3 and DODE routine of Listing 1.4 is considered next.

1.2.3 Numerical, graphical output

We can note the following details about this output.

- 3001 t output points as the first dimension of the solution matrices yout1, yout2 from dede as programmed in the main program of Listing 1.3.

```
[1] 3001

[1] 3

[1] 3001

[1] 3

ncall = 81364
```

Table 1.3   Abbreviated output from Listings 1.3 and 1.4

- The solution matrices `yout1`, `yout2` returned by `dede` have three elements as a second dimension. The first element is the value of t. The second is $y(t)$ from eq. (1.2-1) (for each of the 3001 output points). The third is $y(t - \tau)$ used in eq. (1.2-1).

- The computational effort is large, `ncall = 81364`, but the solutions are calculated almost immediately reflecting a single DODE and the efficiency of `lagvalue` and `dede`.

The graphical output is in Figure 1.2-1 (the complicated variation in t indicates why 3001 points are used to define the solutions).

Figure 1.2-1 indicates the solution $y(t)$ and lag variable $y(t - \tau)$ have two distinct forms:

- For $\tau = 10$, the solution is deterministic and oscillatory. Thus, the lag variable $y(t - 10) =$ `ylag` (plot(row=1,col=2)) has a limit cycle reflecting the oscillation of $y(t)$ (plot (1,1)).

- For $\tau = 20$, the solution is chaotic as reflected in the solution $y(t)$ (plot(2,1)) and the lag variable $y(t - 20)$ (plot(2,2)).

In summary, eqs. (1.2) demonstrate the sensitivity of the solution to lag τ, which can be readily examined with the DODE routines `dede`, `lagvalue`. In particular, Figure 1.2-1 indicates that the solution does not approach a single equilibrium solution value.

Eqs. (1.2) have been reported as a model for the dynamics of WBC [2, p. 125]. Cases of interest include the reduction of the rate constant b to represent the reduction in WBC count resulting from chemotherapy. The study of this application of the model of eqs. (1.2) is left as an exercise.

The next example demonstrates the variation in the equilibrium solution of a DODE with a variation of the parameters.

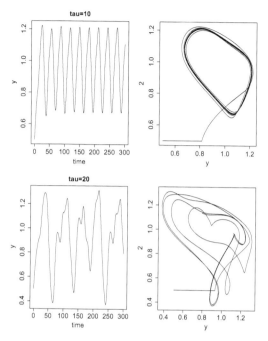

Figure 1.2-1 Numerical solution $y(t)$ and lag variable $y(t - \tau), \tau = 10, 20$ from eqs. (1.2).

1.3   Example 3: One DODE, equilibrium solutions

The following DODE is from [1, p. 238].

$$\frac{dy(t)}{dt} = by(t - 7)\left[1 - y(t)\right] - cy(t); \quad y(t) = \alpha \text{ for } t \leq 0 \qquad (1.3\text{-}1,2)$$

with $0 < \alpha < 1$ in the history, eq. (1.3-2).

For all b, c, $y(t) = 0$ is an equilibrium (steady state) solution to eqs. (1.3). For $b > c$, eqs. (1.3) have a second equilibrium solution, $y(t) = 1 - c/b$ which is demonstrated with the following R routines.

1.3.1   Main program

The following main program is similar to the main programs in Listings 1.1 and 1.3, but some of the discussion is repeated here so that this example (eqs. (1.3)) is self-contained.

```
#
# Epidemic model
#
# Delete previous workspaces
  rm(list=ls(all=TRUE))
#
# Access DODE integrator
  library(deSolve)
#
# Access functions for numerical solution
  setwd("f:/dpde/chap1/ex3");
  source("ode1a.R");
#
# Parameters
  alpha=0.8;
  b=2;
  c=1;
  lag=7;
#
# Temporal grid
  nout=31;t0=0;tf=60;
  times=seq(from=t0,to=tf,by=(tf-t0)/(nout-1));
#
# IC vector
  y0=rep(0,1);
  y0[1]=0.8;
  ncall=0;
#
# Integration of DODE
  yout=dede(y=y0,times=times,func=ode1a);
  nrow(yout);
  ncol(yout);
#
# Display numerical solution
  cat(sprintf("\n        t      y(t)\n"));
  for(it in 1:nout){
    cat(sprintf("\n %6.2f %8.4f",yout[it,1],yout[it,2]));
  }
#
# Arrays for plotted solution
  tp=rep(0,nout);
  yp=rep(0,nout);
  for(it in 1:nout){
    tp[it]=yout[it,1];
    yp[it]=yout[it,2];
  }
#
# Display calls to ode1a
  cat(sprintf("\n ncall = %4d\n",ncall));
```

```
#
# Plotted solution
  par(mfrow=c(1,1));
  matplot(tp,yp,type="l",xlab="t",ylab="y(t)",
    lty=1,main="",lwd=2,col="black");
```

Listing 1.5 Main program for eqs. (1.3).

We can note the following details about Listing 1.5.

- Previous workspaces are deleted.

```
#
# Epidemic model
#
# Delete previous workspaces
  rm(list=ls(all=TRUE))
```

- The R ODE integrator library deSolve is accessed. Then the directory with the files for the solution of eqs. (1.3) is designated. Note that setwd uses / rather than the usual \.

```
#
# Access DODE integrator
  library(deSolve)
#
# Access functions for numerical solution
  setwd("f:/dpde/chap1/ex3");
  source("ode1a.R");
```

ode1a.R is the routine for eqs. (1.3-1,2) (discussed subsequently).

- The parameters in eqs. (1.3) are defined.

```
#
# Parameters
  alpha=0.8;
  b=2;
  c=1;
  lag=7;
```

- An interval in t of 31 points is defined for $0 \leq t \leq 60$ so that tout$=0,2,\ldots,60$.

```
#
# Temporal grid
  nout=31;t0=0;tf=60;
  times=seq(from=t0,to=tf,by=(tf-t0)/(nout-1));
```

- The history (IC) for eq. (1.3-1) is specified as eq. (1.3-2).

```
#
#  IC vector
   y0=rep(0,1);
   y0[1]=0.8;
   ncall=0;
```

Also, the counter for the number of calls to ode1a is initialized.

- Eq. (1.3-1) is integrated by the library integrator dede (available in deSolve, [2, Chapter 7]). As expected, the inputs to dede are the IC vector y0, the vector of output values of t, times, and the DODE function. The length of u0 (1) informs dede how many DODEs are to be integrated. y,times,func are reserved names so the three arguments can be placed in any order in the call to dede.

```
#
#  Integration of DODE
   yout=dede(y=y0,times=times,func=ode1a);
   nrow(yout);
   ncol(yout);
```

The numerical solution to eq. (1.3-1) is returned in matrix yout. In this case, yout has the dimensions *nout* $\times (1+1) = 31 \times 1 + 1 = 2$, which are confirmed by the output from nrow(yout),ncol(yout) (included in the numerical output considered subsequently).

The offset $1 + 1$ is required since the first element of each of the 31 rows has the output t (also in times), and the second element has the solution of eq. (1.3-1).

- The solution $y(t)$ of eq. (1.3-1) is displayed as a function of t (with index it).

```
#
#  Display numerical solution
   cat(sprintf("\n        t        y(t)\n"));
   for(it in 1:nout){
     cat(sprintf("\n %6.2f %8.4f",yout[it,1],yout[it,2]));
   }
```

- The solution $y(t)$ is placed in arrays tp,yp for subsequent plotting.

```
#
#  Arrays for plotted solution
   tp=rep(0,nout);
   yp=rep(0,nout);
   for(it in 1:nout){
     tp[it]=yout[it,1];
     yp[it]=yout[it,2];
   }
```

- The number of calls to `ode1a` is displayed at the end of the solution.

```
#
# Display calls to ode1a
  cat(sprintf("\n ncall = %4d\n",ncall));
```

- The numerical solution to eq. (1.3-1) is plotted against *t*.

```
#
# Plotted solution
  par(mfrow=c(1,1));
  matplot(tp,yp,type="l",xlab="t",ylab="y(t)",
    lty=1,main="",lwd=2,col="black");
```

This completes the discussion of the main program of Listing 1.5. The DODE routine `ode1a` is considered next.

1.3.2   DODE routine

`ode1a` follows.

```
  ode1a=function(t,y,parm,tau) {
#
# DODE source term
  if (t > lag){
    ylag=lagvalue(t-lag);
  } else {
    ylag=alpha;
  }
#
# DODE
  yt=rep(0,1);
  yt[1]=b*ylag[1]*(1-y[1])-c*y[1];
#
# Increment calls to ode1a
  ncall <<- ncall+1;
#
# Return DODE t derivative vector
  return(list(c(yt[1])));
#
# End of ode1a
  }
```

Listing 1.6   DODE routine for eqs. (1.3).

We can note the following details about Listing 1.6.

- The function is defined.

```
  ode1a=function(t,y,parm,tau) {
```

t is the current value of *t* in eqs. (1.3). y is the current numerical solution to eq. (1.3-1). parm is an argument to pass parameters to ode1a (unused, but required in the argument list). tau is the lag (delay) of an DODE system, which is unused in the current application (the lag is passed to ode1a from the main program of Listing 1.5 as parameter lag). The arguments must be listed in the order stated to properly interface with dede called in the main program of Listing 1.5. The derivative $dy(t)/dt$ of eq. (1.3-1) is calculated and returned to dede as explained subsequently.

- The delay term $y(t - \tau)$ in eq. (1.3-1) is programmed.

```
#
# DODE source term
  if (t > lag){
    ylag=lagvalue(t-lag);
  } else {
    ylag=alpha;
  }
```

lagvalue is a R utility that defines past values of y (the second input argument of ode1a). ylag is a vector with the past values of y that can be used in the following programming of eq. (1.3-1).

- Eq. (1.3-1) is programmed. The derivative $dy(t)/dt$ is placed in yt[1].

```
#
# DODE
  yt=rep(0,1);
  yt[1]=b*ylag[1]*(1-y[1])-c*y[1];
```

- The counter for the calls to ode1a is incremented and returned to the main program of Listing 1.5 by «-.

```
#
# Increment calls to ode1a
  ncall <<- ncall+1;
```

- The derivative yt is returned to dede for the next step along the solution.

```
#
# Return DODE t derivative vector
  return(list(c(yt[1])));
```

yt is returned as a list as required by dede. c is the R vector utility.

- The final } concludes ode1a.

```
#
# End of ode1a
  }
```

The numerical and graphical (plotted) output from the main program of Listing 1.5 and DODE routine of Listing 1.6 is considered next.

1.3.3 Numerical, graphical output

The numerical output follows in Table 1.4.

```
[1] 31

[1] 2

      t       y(t)
   0.00    0.8000
   2.00    0.6164
   4.00    0.6154
   6.00    0.6154
   8.00    0.5720
  10.00    0.5522
      .         .
      .         .
      .         .
Output for t = 12
   to 48 removed
      .         .
      .         .
      .         .
  50.00    0.5015
  52.00    0.5013
  54.00    0.5009
  56.00    0.5007
  58.00    0.5007
  60.00    0.5006

ncall = 1153
```

Table 1.4   Abbreviated output from Listings 1.5 and 1.6

We can note the following details about this output.

- 31 t output points as the first dimension of the solution matrix yout from dede as programmed in the main program of Listing 1.5.

- The solution matrix yout returned by dede has two elements as a second dimension. The first element is the value of t. The second is $y(t)$ from eq. (1.3-1) (for each of the 31 output points).

- The solution approaches the equilibrium value $1 - c/b = 1 - 1/2 = 0.5$ as explained previously.

- The computational effort is modest, `ncall = 1153`, indicating `dede` calculates the solution efficiently.

The graphical output is in Figure 1.3-1. The nonsmooth variation in t reflects the effect of the delay `lag=7`. This could be studied further by increasing the number of output points (above `31`). This is left as an exercise. Also, the case $b < c$ can be studied (the steady state solution is $y(t) = 0$).

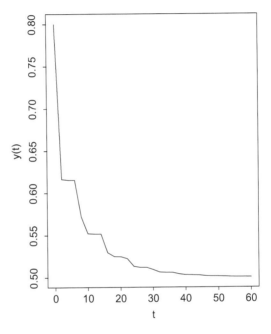

Figure 1.3-1   Numerical solution $y(t)$ from eqs. (1.3).

Figure 1.3-1 indicates the solution $y(t)$ approaches 0.5 as a steady state solution.

In summary, the solution of DODE eq. (1.3-1) with history (1.3-2) confirms the expected single-value equilibrium solutions.

The next example is an extension of one DODE to simultaneous DODEs.

1.4   Example 4: Simultaneous DODEs

The following 3×3 DODE system (three DODEs in three unknowns) is reported as an epidemiological model ([2, pp. 133–134]).

$$\frac{dy_1}{dt} = -y_1(t)y_2(t-1) + y_2(t-10) \tag{1.4-1}$$

$$\frac{dy_2}{dt} = y_1(t)y_2(t-1) - y_2(t) \tag{1.4-2}$$

$$\frac{dy_3}{dt} = y_2(t) - y_2(t-10) \tag{1.4-3}$$

The history for eqs. (1.4) is

$$y(t) = [5, 0.1, 1]^T \text{ for } t \le 0 \tag{1.4-4}$$

Eqs. (1.4) constitute a simultaneous, nonlinear DODE system with two delays. Eqs. (1.4) also demonstrate the efficacy of a numerical solution since an analytical solution is probably unavailable. In order to validate the numerical solution, a reference solution has been reported ([2, p. 133]).

$$y_1(t = 40) = 0.0912491205663460$$
$$y_2(t = 40) = 0.0202995003350707$$
$$y_3(t = 40) = 5.98845137909849 \tag{1.4-5}$$

A main program and DODE routine for eqs. (1.4) follows.

1.4.1   Main program

The following main program is similar to the main programs in Listings 1.1, 1.3, and 1.5, but some of the discussion is repeated here so that this example (eqs. (1.4)) is self-contained.

```
#
# Epidemic model, three DODEs
#
# Delete previous workspaces
  rm(list=ls(all=TRUE))
#
# Access DODE integrator
  library(deSolve)
#
# Access functions for numerical solution
  setwd("f:/dpde/chap1/ex4");
  source("ode1a.R");
#
# Parameters
  tau1=1;
  tau2=10;
#
# Temporal grid
  nout=401;t0=0;tf=40;
  times=seq(from=t0,to=tf,by=(tf-t0)/(nout-1));
#
# IC vector
  y0=rep(0,3);
```

```
  y0[1]=5;
  y0[2]=0.1;
  y0[3]=1;
  ncall=0;
#
# Integration of DODE
  yout=dede(y=y0,times=times,func=ode1a);
  nrow(yout);
  ncol(yout);
#
# Vectors for DODE solution
  y1p=rep(0,nout);
  y2p=rep(0,nout);
  y3p=rep(0,nout);
  tp=rep(0,nout);
  for(it in 1:nout){
     tp[it]=yout[it,1];
    y1p[it]=yout[it,2];
    y2p[it]=yout[it,3];
    y3p[it]=yout[it,4];
  }
#
# Display numerical solution
  cat(sprintf("\n    it       t     y1(t)     y2(t)      y3(t)\n"));
  iv=seq(from=1,to=nout,by=10);
  for(it in iv){
    cat(sprintf("\n %4d %6.2f %8.4f %8.4f %8.4f",
                    it,tp[it],y1p[it],y2p[it],y3p[it]));}
#
# Display calls to ode1a
  cat(sprintf("\n ncall = %4d\n",ncall));
#
# Plot DODE solution against t
  par(mfrow=c(1,1));
  plot(tp,y1p,type="l",xlab="t",ylab="y1(t)",
       lty=1,main="y1(t)",lwd=2,col="black");
  plot(tp,y2p,type="l",xlab="t",ylab="y2(t)",
       lty=1,main="y2(t)",lwd=2,col="black");
  plot(tp,y3p,type="l",xlab="t",ylab="y3(t)",
       lty=1,main="y3(t)",lwd=2,col="black");
```

Listing 1.7   Main program for eqs. (1.4).

We can note the following details about Listing 1.7.

- Previous workspaces are deleted.

```
  #
  # Epidemic model, three DODEs
  #
  # Delete previous workspaces
    rm(list=ls(all=TRUE))
```

- The R ODE integrator library deSolve is accessed. Then the directory with the files for the solution of eqs. (1.4) is designated. Note that setwd uses / rather than the usual \.

```
#
# Access DODE integrator
  library(deSolve)
#
# Access functions for numerical solution
  setwd("f:/dpde/chap1/ex4");
  source("ode1a.R");
```

ode1a.R is the routine for eqs. (1.4-1,2,3) (discussed subsequently).

- The parameters in eqs. (1.4-1,2,3) are defined.

```
#
# Parameters
  tau1=1;
  tau2=10;
```

- An interval in t of 401 points is defined for $0 \le t \le 40$ so that tout=0,0.1,...,40.

```
#
# Temporal grid
  nout=401;t0=0;tf=40;
  times=seq(from=t0,to=tf,by=(tf-t0)/(nout-1));
```

- The history (IC), eq. (1.4-4), is programmed.

```
#
# IC vector
  y0=rep(0,3);
  y0[1]=5;
  y0[2]=0.1;
  y0[3]=1;
  ncall=0;
```

Also, the counter for the number of calls to ode1a is initialized.

- Eqs. (1.4-1,2,3) are integrated by the library integrator dede (available in deSolve, [2, Chapter 7]). As expected, the inputs to dede are the IC vector y0, the vector of output values of t, times, and the DODE function. The length of u0 (3) informs dede how many DODEs are to be integrated. y, times, func are reserved names so the three arguments can be placed in any order in the call to dede.

```
#
# Integration of DODE
  yout=dede(y=y0,times=times,func=ode1a);
  nrow(yout);
  ncol(yout);
```

The numerical solution to eqs. (1.4-1,2,3) is returned in matrix yout. In this case, yout has the dimensions $nout \times (3+1) = 401 \times 3 + 1 = 4$, which are confirmed by the output from nrow(yout),ncol(yout) (included in the numerical output considered subsequently).

The offset $+1$ is required since the first element of each of the 401 rows has the output t (also in times), and the second to fourth elements have the solutions of eqs. (1.4-1,2,3).

- The solutions $y_1(t), y_2(t), y_3(t)$ are placed in arrays tp,yp1,yp2,yp3 for subsequent plotting.

```
#
# Vectors for DODE solution
  y1p=rep(0,nout);
  y2p=rep(0,nout);
  y3p=rep(0,nout);
  tp=rep(0,nout);
  for(it in 1:nout){
     tp[it]=yout[it,1];
    y1p[it]=yout[it,2];
    y2p[it]=yout[it,3];
    y3p[it]=yout[it,4];
  }
```

- The numerical solutions to eqs. (1.4-1,2,3) are displayed.

```
#
# Display numerical solution
  cat(sprintf("\n   it     t    y1(t)     y2(t)     y3(t)\n"));
  iv=seq(from=1,to=nout,by=10);
  for(it in iv){
    cat(sprintf("\n %4d %6.2f %8.4f %8.4f %8.4f",
                    it,tp[it],y1p[it],y2p[it],y3p[it]));}
```

Every tenth value in t is displayed with by=10.

- The number of calls to ode1a is displayed at the end of the solution.

```
#
# Display calls to ode1a
  cat(sprintf("\n ncall = %4d\n",ncall));
```

- The numerical solutions to eqs. (1.4-1,2,3) are plotted against t.

```
#
# Plot DODE solution against t
  par(mfrow=c(1,1));
  plot(tp,y1p,type="l",xlab="t",ylab="y1(t)",
       lty=1,main="y1(t)",lwd=2,col="black");
  plot(tp,y2p,type="l",xlab="t",ylab="y2(t)",
       lty=1,main="y2(t)",lwd=2,col="black");
  plot(tp,y3p,type="l",xlab="t",ylab="y3(t)",
       lty=1,main="y3(t)",lwd=2,col="black");
```

This completes the discussion of the main program of Listing 1.7. The DODE routine ode1a is considered next.

1.4.2 DODE routine

ode1a follows.

```
  ode1a=function(t,y,parm,lag) {
#
# DODE source terms
  if (t > tau1){
    ylag1=lagvalue(t-tau1);
  } else {
    ylag1=y0;
  }
  if (t > tau2){
    ylag2=lagvalue(t-tau2);
  } else {
    ylag2=y0;
  }
#
# DODEs
  yt=rep(0,3);
  yt[1]=-y[1]*ylag1[2]+ylag2[2];
  yt[2]= y[1]*ylag1[2]-y[2];
  yt[3]= y[2]-ylag2[2];
#
# Increment calls to ode1a
  ncall <<- ncall+1;
#
# Return DODE t derivative vector
  return(list(c(yt)));
#
# End of ode1a
  }
```

Listing 1.8 DPDE routine for eqs. (1.4-1,2,3,4).

We can note the following details about Listing 1.8.

- The function is defined.

```
ode1a=function(t,y,parm,lag) {
```

  t is the current value of *t* in eqs. (1.4-1,2,3). y is the current numerical solution to eqs. (1.4-1,2,3). parm is an argument to pass parameters to pde1a (unused, but required in the argument list). lag is the lag (delay) of an DODE system, which is unused in the current application (the lags are passed to ode1a from the main program of Listing 1.7 as parameters tau1, tau2).

  The arguments must be listed in the order stated to properly interface with dede called in the main program of Listing 1.7. The derivatives $dy_1(t)/dt$, $dy_2(t)/dt$, $dy_3(t)/dt$ of eqs. (1.4-1,2,3) are calculated and returned to dede as explained subsequently.

- The delay terms in eqs. (1.4-1,2,3) are programmed.

```
#
# DODE source terms
  if (t > tau1){
    ylag1=lagvalue(t-tau1);
  } else {
    ylag1=y0;
  }
  if (t > tau2){
    ylag2=lagvalue(t-tau2);
  } else {
    ylag2=y0;
  }
```

  lagvalue is a R utility that defines past values of y (the second input argument of ode1a). ylag1 is a vector with the past values for tau=1 that can be used in the following programming of eq. (1.4-1,2,3). ylag2 is a vector with the past values for tau=10. In both cases, for $t \leq 0$, the history (IC) vector defined in the main program of Listing 1.7 is used.

- Eqs. (1.4-1,2,3) are programmed. The three derivatives (LHSs of eqs. (1.4-1,2,3)) are placed in yt[1],yt[2],yt[3].

```
#
# DODEs
  yt=rep(0,3);
  yt[1]=-y[1]*ylag1[2]+ylag2[2];
  yt[2]= y[1]*ylag1[2]-y[2];
  yt[3]= y[2]-ylag2[2];
```

  A comparison with eqs. (1.4-1,2,3) gives an explanation of this code.

- The counter for the calls to ode1a is incremented and returned to the main program of Listing 1.7 by «-.

```
#
# Increment calls to ode1a
  ncall <<- ncall+1;
```

- The derivative yt is returned to dede for the next step along the solution.

```
#
# Return DODE t derivative vector
  return(list(c(yt)));
```

yt is returned as a list as required by dede. c is the R vector utility.

- The final } concludes ode1a.

```
#
# End of ode1a
  }
```

The numerical and graphical (plotted) output from the main program of Listing 1.7 and DODE routine of Listing 1.8 is considered next.

1.4.3 Numerical, graphical output

The numerical output follows in Table 1.5.
We can note the following details about this output.

- 401 t output points as the first dimension of the solution matrix yout from dede as programmed in the main program of Listing 1.7.

- The solution matrix yout returned by dede has four elements as a second dimension. The first element is the value of t. The second to fourth are $y_1(t), y_2(t), y_3(t)$ from eqs. (1.4-1,2,3) (for each of the 401 output points).

- The solutions approach the reference solutions, eqs. (1.4-5).

- The computational effort is manageable, ncall = 5585, indicating dede calculates the solution efficiently.

The graphical output is in Figures 1.4-1,2,3. The three ICs of eqs. (1.4-4) are confirmed, and the three DODE-dependent variables indicate that the modeled epidemic does not fade, but continues to oscillate.

In summary, the solution of eqs. (1.4-1,2,3) with history (1.4-4) demonstrates the numerical solution of a simultaneous, nonlinear, DODE system with multiple delays (two in this case).

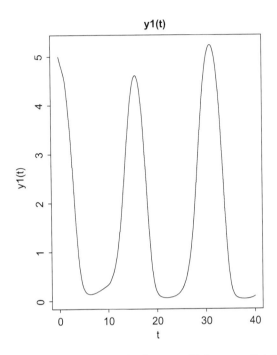

Figure 1.4-1   Numerical solution $y_1(t)$ from eq. (1.4-1).

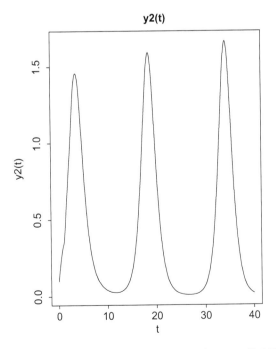

Figure 1.4-2   Numerical solution $y_2(t)$ from eq. (1.4-2).

```
[1] 401

[1] 4

    it      t     y1(t)     y2(t)     y3(t)
     1   0.00    5.0000    0.1000    1.0000
    11   1.00    4.6193    0.3386    1.1420
    21   2.00    3.7136    0.7986    1.5879
    31   3.00    2.2254    1.3302    2.5444
    41   4.00    0.8327    1.3980    3.8693
    51   5.00    0.2534    0.9047    4.9419
    61   6.00    0.1398    0.4567    5.5035
    71   7.00    0.1481    0.2231    5.7288
    81   8.00    0.1940    0.1150    5.7910
    91   9.00    0.2581    0.0643    5.7776
   101  10.00    0.3329    0.0392    5.7280
                   .                   .
                   .                   .
                   .                   .

        Output for t = 11 to 29 removed

                   .                   .
                   .                   .
                   .                   .
   301  30.00    4.8725    0.0733    1.1542
   311  31.00    5.2314    0.1943    0.6743
   321  32.00    4.8648    0.4993    0.7360
   331  33.00    3.6049    1.0962    1.3990
   341  34.00    1.6916    1.6459    2.7625
   351  35.00    0.4229    1.3593    4.3178
   361  36.00    0.0936    0.6883    5.3181
   371  37.00    0.0405    0.2883    5.7712
   381  38.00    0.0352    0.1159    5.9489
   391  39.00    0.0484    0.0471    6.0045
   401  40.00    0.0912    0.0203    5.9885

ncall = 5585
```

Table 1.5   Abbreviated output from Listings 1.7 and 1.8

   To conclude this chapter, the four preceding examples present some basic properties of DODEs and the corresponding numerical solutions produced with R utilities. Examples 2–4 also demonstrate DODE biomedical applications. Additional details about the applications are available from the original sources indicated in [1, 2].

   DODEs also serve as the starting point for the computer-based analysis of delay partial differential equations (DPDEs), which are the mathematical

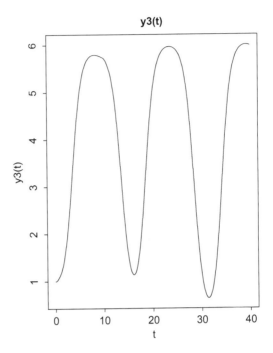

Figure 1.4-3   Numerical solution $y_3(t)$ from eq. (1.4-3).

forms discussed subsequently in this book, again presented through biomedical examples.

References

[1] Shampine, L.F., I. Gladwell, and S. Thompson (2003), *Solving ODEs with Matlab*, Cambridge University Press, Cambridge, UK.

[2] Soetaert, K., J. Cash, and F. Mazzia (2012), *Solving Differential Equations in R*, Springer-Verlag, Heidelberg, Germany.

2

Introduction to Delay Partial Differential Equations

Introduction

In Chapter 1, four delay ordinary differential equation (DODE) examples are discussed which demonstrate: (1) the computer implementation of the DODEs and (2) some properties of the numerical solutions, particularly the effect of the delays. In this chapter, the first DODE example in Chapter 1 is extended to a delay partial differential equation (DPDE). The computer implementation of the DPDE is discussed in detail, and the numerical solutions for several cases are presented.

2.1 DPDE model

The DODE of eqs. (1.1-1,2) is extended to a DPDE by adding a spatial independent variable, x. The equations are

$$\frac{\partial u(x,t)}{\partial t} = d\frac{\partial^2 u(x,t)}{\partial x^2} - au(x,t-1) - bu(x,t) \qquad (2.1\text{-}1)$$

where[1]

Variable, parameter	Explanation, interpretation
$u(x,t)$	DPDE dependent variable
$u(x,t-1)$	delayed dependent variable
x	spatial independent variable
t	temporal (initial value) independent variable
d	diffusivity
a,b	constants

Table 2.1 Variables and parameters of eq. (2.1-1)

---

[1]Following the usual convention of the numerical analysis literature, the dependent variable of DODE (1.1-1) is designated $y(t)$, and of DPDE (2.1-1), $u(x,t)$ (y and u for ODEs and PDEs, respectively).

For the special case $d = b = 0$ eq. (2.1-1) reduces to eq. (1.1-1), and the solutions are subsequently compared as a test of the coding (programming) of eq. (2.1-1).

Eq. (2.1-1) is second order in x and first order in t, so it requires two boundary conditions (BCs) and one initial condition (IC).

$$\frac{\partial u(x = x_l, t)}{\partial x} = \frac{\partial u(x = x_u, t)}{\partial x} = 0 \qquad (2.1\text{-}2,3)$$

$$u(x, t = 0) = u_0(x) \qquad (2.1\text{-}4)$$

As in the case of eq. (1.1-1), the history vector of eq. (2.1-1) is taken as $u_0(x) = 1$.

Eqs. (2.1) are coded in a set of R routines, starting with the main program.

2.1.1 Main program

A main program for eqs. (2.1) follows.

```
#
# Basic DPDE
#
# Delete previous workspaces
  rm(list=ls(all=TRUE))
#
# Access DODE integrator
  library(deSolve)
#
# Access functions for numerical solution
  setwd("f:/dpde/chap2");
  source("pde1a.R");
  source("dss004.R");
  source("dss044.R");
#
# Model parameters
  tau=1;
  a=1;
  b=0;
  d=0;
#
# Spatial grid
  nx=21;xl=0;xu=1;
  x=seq(from=xl,to=xu,by=(xu-xl)/(nx-1));
#
# Temporal grid
  nout=21;t0=0;tf=10
  times=seq(from=t0,to=tf,by=(tf-t0)/(nout-1));
#
# IC vector
```

```
  u0=rep(0,nx);
  for(i in 1:nx){
    u0[i]=1;
  }
  ncall=0;
#
# Integration of delay MOL/ODEs
  uout=dede(y=u0,times=times,func=pde1a);
  nrow(uout);
  ncol(uout);
#
# Arrays/vectors for ODE solutions
  up=matrix(0,nrow=nx,ncol=nout);
  tp=rep(0,nout);
  for(it in 1:nout){
    tp[it]=uout[it,1];
  for( i in 1:nx){
    up[i,it]=uout[it,i+1];
  }
  }
#
# Display numerical solution
  iv=seq(from=1,to=nout,by=5);
  for(it in iv){
  cat(sprintf(
    "\n\n        t             x       u(x,t)"));
  iv=seq(from=1,to=nx,by=5);
  for(i in iv){
    cat(sprintf("\n%9.2e%11.2e%12.4f",
      tp[it],x[i],up[i,it]));
  }
  }
#
# Display calls to ode1a
  cat(sprintf("\n ncall = %4d\n",ncall));
#
# Plot MOL/ODE solutions
  par(mfrow=c(1,1));
#
# 2D
  matplot(x,up,type="l",xlab="x",ylab="u(x,t)",
          lty=1,main="",lwd=2,col="black");
#
# 3D
  persp(x,tp,up,theta=60,phi=30,
        xlim=c(xl,xu),ylim=c(t0,tf),xlab="x",
        ylab="t",zlab="u(x,t)");
```

Listing 2.1   Main program for eqs. (2.1).

The following discussion of Listing 2.1 partly repeats the discussion of Listing 1.1 in Chapter 1, but is given so that this first DPDE example (eqs. (2.1)) is self-contained.

- Previous workspaces are deleted.

```
#
# Basic DPDE
#
# Delete previous workspaces
  rm(list=ls(all=TRUE))
```

- The R ODE integrator library deSolve is accessed. Then the directory with the files for the solution of eqs. (2.1) is designated. Note that setwd (set working directory) uses / rather than the usual \.

```
#
# Access DODE integrator
  library(deSolve)
#
# Access functions for numerical solution
  setwd("f:/dpde/chap2");
  source("pde1a.R");
  source("dss004.R");
  source("dss044.R");
```

pde1a.R is the routine for eqs. (2.1) (discussed subsequently) based on the method of lines (MOL), a general algorithm for PDEs [1]. dss004, dss044 are library routines for the calculation of first and second spatial derivatives. These routines are listed in Appendix A1 with additional explanation.

- The parameters (constants) of eq. (2.1-1) are specified.

```
#
# Model parameters
  tau=1;
  a=1;
  b=0;
  d=0;
```

For this first case, with b=d=0, eq. (2.1-1) reduces to eq. (1.1-1) as demonstrated subsequently.

- A spatial grid is defined for nx=21 points in x, with $x_l = 0, x_u = 1$ in BCs (2.1-2,3), so $x = 0, 0.05, ..., 1$ (from the seq).

```
#
# Spatial grid
  nx=21;xl=0;xu=1;
  x=seq(from=xl,to=xu,by=(xu-xl)/(nx-1));
```

- A temporal interval is defined with `nout=21` output points in t, initial and final values of `t0=0`, `tf=10`, so that $t = 0, 0.5, ..., 10$.

```
#
# Temporal grid
  nout=21;t0=0;tf=10
  times=seq(from=t0,to=tf,by=(tf-t0)/(nout-1));
```

- An IC and history vector is defined for `nx=21` points in x (rather than just one value as for eqs. (1.1-1,2)).

```
#
# IC vector
  u0=rep(0,nx);
  for(i in 1:nx){
    u0[i]=1;
  }
  ncall=0;
```

The counter for the calls to `pde1a` is also initialized.

- Eq. (2.1-1) is integrated (solved numerically) by the library integrator dede (available in `deSolve`, [3, Chapter 7]). As expected, the inputs to dede are the IC vector `u0`, the vector of output values of t, `times`, and the ODE function, `pde1a`. The length of `u0` (21) informs dede how many ODEs are to be integrated. `y`, `times`, `func` are reserved names.

```
#
# Integration of delay MOL/ODEs
  uout=dede(y=u0,times=times,func=pde1a);
  nrow(uout);
  ncol(uout);
```

The numerical solution to eq. (2.1-1) is returned in matrix `uout`. In this case, `uout` has the dimensions $nout = 21 \times nx + 1 = 21 + 1 = 22$, which are confirmed by the output from `nrow(out),ncol(out)` (included in the numerical output considered subsequently).

- An array, `up`, and a vector, `tp`, are defined for the numerical solution $u(x,t)$ and t from dede (`uout`).

```
#
# Arrays/vectors for ODE solutions
  up=matrix(0,nrow=nx,ncol=nout);
  tp=rep(0,nout);
  for(it in 1:nout){
    tp[it]=uout[it,1];
  for( i in 1:nx){
    up[i,it]=uout[it,i+1];
  }
  }
```

- The solution to eq. (2.1-1), $u(x,t)$, is displayed as a function of x and t with two `for`s.

```
#
# Display numerical solution
  iv=seq(from=1,to=nout,by=5);
  for(it in iv){
  cat(sprintf(
    "\n\n            t          x        u(x,t)"));
  iv=seq(from=1,to=nx,by=5);
  for(i in iv){
    cat(sprintf("\n%9.2e%11.2e%12.4f",
      tp[it],x[i],up[i,it]));
  }
  }
```

Every fifth value of x and t is displayed with `by=5`.

- The number of calls to `pde1a` is displayed at the end of the solution.

```
#
# Display calls to pde1a
  cat(sprintf("\n ncall = %4d\n",ncall));
```

- $u(x,t)$ is plotted against x in two dimension (2D) with `matplot`.

```
#
# Plot MOL/ODE solutions
  par(mfrow=c(1,1));
#
# 2D
  matplot(x,up,type="l",xlab="x",ylab="u(x,t)",
          lty=1,main="",lwd=2,col="black");
```

The plots are parametric in t.

- $u(x,t)$ is plotted against x and t in three dimension (3D) with `persp`.

```
#
# 3D
  persp(x,tp,up,theta=60,phi=30,
        xlim=c(xl,xu),ylim=c(t0,tf),xlab="x",
        ylab="t",zlab="u(x,t)");
```

The angles `theta`, `phi` were selected by trial and error to produce a 3D plot with clear orientation in space (presented subsequently).

This concludes the discussion of the main program in Listing 2.1. The MOL/ODE routine `pde1a` is considered next.

2.1.2 DODE routine

The MOL/ODE routine `pde1a` called by `dede` is listed next.

```
   pde1a=function(t,u,parms,lag) {
#
# DPDE source term
   if (t > tau){
     ulag=lagvalue(t-tau);
   } else {
     ulag=u0;
   }
#
# ux
   ux=dss004(xl,xu,nx,u);
#
# BCs
   ux[1]=0;
   ux[nx]=0;
#
# uxx
   nl=2;nu=2;
   uxx=dss044(xl,xu,nx,u,ux,nl,nu);
#
# DPDE
   ut=rep(0,nx);
   for(i in 1:nx){
     ut[i]=d*uxx[i]-a*ulag[i]-b*u[i];
   }
#
# Increment calls to ncall
   ncall <<- ncall+1;
#
# Return DPDE t vector
   return(list(c(ut)));
#
# End of pde1a
   }
```

Listing 2.2   MOL/ODE routine for eqs. (2.1).

We can note the following details about this listing.

- The function is defined.

```
   pde1a=function(t,u,parm,lag) {
```

t is the current value of t in eqs. (2.1). u is the current numerical solution to eq. (2.1-1). parm is an argument to pass parameters to pde1a (unused, but required in the argument list). lag is the lag (delay) of the DPDE system, which is unused in the current application (the lag is passed to `pde1a` from the main program of Listing 2.1 as parameter `tau`). The arguments must be listed in the order stated to properly interface with `dede` called in the main program of Listing 2.1. The derivative $\dfrac{\partial u}{\partial t}$ of eq. (2.1-1) is calculated and returned to `dede` as explained subsequently.

- The lagged variable $u(t-1)$ in eq. (2.1-1) is computed as `ulag`. The use of the R utility `lagvalue` is discussed after Listing 1.2.

```
#
# DPDE source term
  if (t > tau){
    ulag=lagvalue(t-tau);
  } else {
    ulag=u0;
  }
```

Note the use of the history vector `u0`, and the lag (delay) `tau`.

- The first derivative $\dfrac{\partial u}{\partial x}$ is computed with `dss004`. The arguments of `dss004` are explained in Appendix A1.

```
#
# ux
  ux=dss004(xl,xu,nx,u);
```

- BCs (2.1-2,3) are implemented.

```
#
# BCs
  ux[1]=0;
  ux[nx]=0;
```

Subscripts `1,nx` correspond to the boundary values of $\dfrac{\partial u}{\partial x}$ at $x = x_l = 0, x = x_u = 1$, respectively.

- The second derivative $\dfrac{\partial^2 u}{\partial x^2}$ in eq. (2.1-1) is computed with `dss044` (listed and discussed in Appendix A1).

```
#
# uxx
  nl=2;nu=2;
  uxx=dss044(xl,xu,nx,u,ux,nl,nu);
```

`nl=nu=2` specify Neumann BCs (eqs. (2.1-2,3)). The array `ux` computed previously provides the boundary values to `dss044`.

- The MOL programming of eq. (2.1-1) steps through the 21 values of x in a `for`.

```
#
# DPDE
  ut=rep(0,nx);
  for(i in 1:nx){
    ut[i]=d*uxx[i]-a*ulag[i]-b*u[i];
  }
```

The correspondence of the DPDE (eq. (2.1-1)) and the programming is a principal feature of the MOL.

- The counter for the calls to `pde1a` is incremented and returned to the main program of Listing 2.1 by «-.

```
#
# Increment calls to ncall
  ncall <<- ncall+1;
```

- The derivative vector `ut` is returned to `dede` for the next step along the solution.

-
```
#
# Return DPDE t vector
  return(list(c(ut)));
#
# End of pde1a
  }
```

The derivative `ut` is returned as a `list` as required by `dede`. c is the R vector utility. The final } concludes `pde1a`.

This completes the discussion of the MOL/ODE routine `pde1a`. The output from the main program of Listing 2.1 and MOL/ODE routine of Listing 2.2 is considered next.

2.1.3 Numerical, graphical output, no diffusion

Abbreviated numerical output is in Table 2.2.

```
[1] 21

[1] 22
```

t	x	u(x,t)
0.00e+00	0.00e+00	1.0000
0.00e+00	2.50e-01	1.0000
0.00e+00	5.00e-01	1.0000

```
0.00e+00    7.50e-01        1.0000
0.00e+00    1.00e+00        1.0000

     t            x          u(x,t)
2.50e+00    0.00e+00       -0.3958
2.50e+00    2.50e-01       -0.3958
2.50e+00    5.00e-01       -0.3958
2.50e+00    7.50e-01       -0.3958
2.50e+00    1.00e+00       -0.3958

     t            x          u(x,t)
5.00e+00    0.00e+00        0.1583
5.00e+00    2.50e-01        0.1583
5.00e+00    5.00e-01        0.1583
5.00e+00    7.50e-01        0.1583
5.00e+00    1.00e+00        0.1583

     t            x          u(x,t)
7.50e+00    0.00e+00       -0.0594
7.50e+00    2.50e-01       -0.0594
7.50e+00    5.00e-01       -0.0594
7.50e+00    7.50e-01       -0.0594
7.50e+00    1.00e+00       -0.0594

     t            x          u(x,t)
1.00e+01    0.00e+00        0.0202
1.00e+01    2.50e-01        0.0202
1.00e+01    5.00e-01        0.0202
1.00e+01    7.50e-01        0.0202
1.00e+01    1.00e+00        0.0202

ncall =   409
```

Table 2.2   Abbreviated output from Listings 2.1 and 2.2

We can note the following details about this output.

- 21 t output points as the first dimension of the solution matrix uout from dede as programmed in the main program of Listing 2.1.

- The solution matrix uout returned by dede has 22 elements as a second dimension. The first element is the value of t. Elements 2–22 are $u(x,t)$ from eq. (2.1-1) (for each of the 21 output points).

- The solution is displayed for $t = 0, 2.5, ..., 10$ and $x = 0, 0.25, ..., 1$ as programmed in Listing 2.1.

- The solution is in agreement with the solution from Table 1.1. For example, at $t = 0$

```
Table 1.1

    it      t      y(t)      ya(t)      diff(t)
    1     0.00    1.000     1.000    0.000e+00
```

```
Table 2.2

        t            x          u(x,t)
   0.00e+00    0.00e+00        1.0000
   0.00e+00    2.50e-01        1.0000
   0.00e+00    5.00e-01        1.0000
   0.00e+00    7.50e-01        1.0000
   0.00e+00    1.00e+00        1.0000
```

so the ICs for eqs. (1.1-1) and (2.1-1) are confirmed.
At $t = 2.5$,

```
Table 1.1

    it      t      y(t)      ya(t)      diff(t)
    26    2.50   -0.396    -0.396   -3.873e-06
```

```
Table 2.2

        t            x          u(x,t)
   2.50e+00    0.00e+00       -0.3958
   2.50e+00    2.50e-01       -0.3958
   2.50e+00    5.00e-01       -0.3958
   2.50e+00    7.50e-01       -0.3958
   2.50e+00    1.00e+00       -0.3958
```

At $t = 10$,

```
Table 1.1

    it      t      y(t)
    101   10.00    0.020
```

```
Table 2.2

        t            x          u(x,t)
   1.00e+01    0.00e+00        0.0202
   1.00e+01    2.50e-01        0.0202
   1.00e+01    5.00e-01        0.0202
   1.00e+01    7.50e-01        0.0202
   1.00e+01    1.00e+00        0.0202
```

This agreement is expected since for $d = b = 0$, eq. (2.1-1) (over the interval $0 \leq x \leq 1$) reduces to eq. (1.1-1).

- Since with $d = 0$ (no diffusion in eq. (2.1-1)) and homogeneous Neumann BCs (2.1-2,3), there are no variations of the solution with x, that is, the profiles of $u(x, t)$ in x are flat. This feature of the solution of eq. (2.1-1) is clear in the graphical output that follows. This case is worth considering since if the solution varied with x, a programming error would be indicated.

- The computational effort is modest, `ncall` $=$ `409`, so that `dede` computed a solution to eqs. (2.1) efficiently and accurately.

The graphical output is in Figures 2.1.

This 2D plot of $u(x, t)$ against x (Figure 2.1-1) demonstrates the flat profiles in x as discussed previously. Note also the consistency of the homogeneous Neumann BCs (2.1-2,3) with the solution.

This 3D plot clarifies the solution of Figure 2.1-1. As discussed previously, the solutions are the same as in Figure 1.1-1.

This completes the discussion of the case $d = b = 0, a = 1$ for a uniform IC. Variations in this case are now considered.

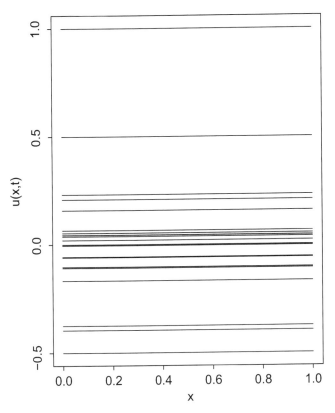

Figure 2.1-1   Numerical solution $u(x, t)$ from eq. (2.1-1), $d = 0$, `matplot`.

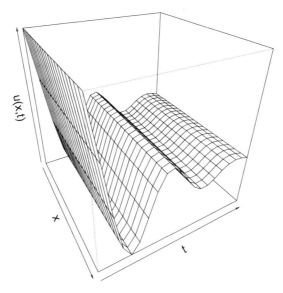

Figure 2.1-2 Numerical solution $u(x,t)$ from eq. (2.1-1), $d = 0$, persp.

If the parameters include a nonzero value of d, diffusion is added to the solution of eq. (2.1-1).

```
#
# Model parameters
  tau=1;
  a=1;
  b=0;
  d=0.05;
```

Listing 2.3 Parameters for eq. (2.1-1) with diffusion.

This choice of d is explained subsequently.
The abbreviated output is given in Table 2.3.

```
[1]  21

[1]  22
```

t	x	u(x,t)
0.00e+00	0.00e+00	1.0000
0.00e+00	2.50e-01	1.0000
0.00e+00	5.00e-01	1.0000
0.00e+00	7.50e-01	1.0000
0.00e+00	1.00e+00	1.0000

t	x	u(x,t)
2.50e+00	0.00e+00	-0.3958
2.50e+00	2.50e-01	-0.3958

```
2.50e+00    5.00e-01      -0.3958
2.50e+00    7.50e-01      -0.3958
2.50e+00    1.00e+00      -0.3958
        .                    .
        .                    .
        .                    .

   Output for t = 5,  7.5 removed
        .                    .
        .                    .

        .                    .
        t           x        u(x,t)
1.00e+01    0.00e+00       0.0202
1.00e+01    2.50e-01       0.0202
1.00e+01    5.00e-01       0.0202
1.00e+01    7.50e-01       0.0202
1.00e+01    1.00e+00       0.0202

ncall = 1223
```

Table 2.3   Abbreviated output from Listings 2.1 and 2.2, $d = 0.05$

The output in Table 2.3 is the same as in Table 2.1 since adding diffusion does not cause the solution to depart from the uniform (in x) IC. The substantial increase in the computation (ncall = 1223 rather than ncall = 409 for no diffusion) results from the effect of the added diffusion.

Again, this case is worth considering since any change in the solution from the case of no diffusion (Table 2.1) would indicate a programming error.

2.1.4   Numerical, graphical output, Dirichlet BCs

If homogeneous Dirichlet BCs,

$$u(x = x_l, t) = u(x = x_u, t) = 0 \qquad (2.1\text{-}5,6)$$

are considered, the programming in pde1a changes.

```
#
# BCs
  u[1]=0;
  u[nx]=0;
#
# uxx
  nl=1;nu=1;
  ux=rep(0,nx);
  uxx=dss044(xl,xu,nx,u,ux,nl,nu);
```
Listing 2.4   Programming for homogeneous Dirichlet BCs (2.1-5,6).

The numerical output is in Table 2.4.

```
[1]  21

[1]  22
```

t	x	u(x,t)
0.00e+00	0.00e+00	1.0000
0.00e+00	2.50e-01	1.0000
0.00e+00	5.00e-01	1.0000
0.00e+00	7.50e-01	1.0000
0.00e+00	1.00e+00	1.0000

t	x	u(x,t)
2.50e+00	0.00e+00	-0.0004
2.50e+00	2.50e-01	-0.1088
2.50e+00	5.00e-01	-0.1746
2.50e+00	7.50e-01	-0.1088
2.50e+00	1.00e+00	-0.0004

t	x	u(x,t)
5.00e+00	0.00e+00	-0.0000
5.00e+00	2.50e-01	-0.0315
5.00e+00	5.00e-01	-0.0452
5.00e+00	7.50e-01	-0.0315
5.00e+00	1.00e+00	-0.0000

t	x	u(x,t)
7.50e+00	0.00e+00	-0.0000
7.50e+00	2.50e-01	0.0249
7.50e+00	5.00e-01	0.0351
7.50e+00	7.50e-01	0.0249
7.50e+00	1.00e+00	-0.0000

t	x	u(x,t)
1.00e+01	0.00e+00	0.0000
1.00e+01	2.50e-01	-0.0074
1.00e+01	5.00e-01	-0.0104
1.00e+01	7.50e-01	-0.0074
1.00e+01	1.00e+00	0.0000

```
ncall = 2073
```

Table 2.4 Abbreviated output from Listings 2.1 and 2.4, $d = 0.05$, BCs (2.1-5,6)

The solution confirms BCs (2.1-5,6) and indicates a maximum or minimum at the midpoint $x = (x_u - x_l)/2 = 0.5$ (depending on the value of t). The computational effort is modest at ncall = 2073.

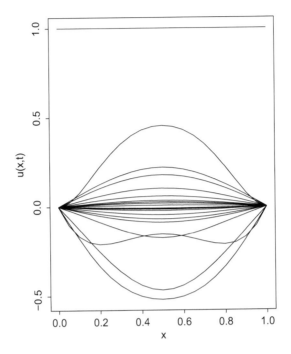

Figure 2.2-1 Numerical solution $u(x,t)$ from eq. (2.1-1), $d = 0.05$, BCs (2.1-5,6), `matplot`.

This 2D plot of $u(x,t)$ against x demonstrates the variation with x. Note that BCs (2.1-5,6) introduce a discontinuous change at $x = x_l, x_u$ from the initial values $u(x = x_l, x_u, t = 0) = 1$ to the boundary values $u(x = x_l, x_u, t > 0) = 0$. For large t, the solution approaches $u(x, t \to \infty) = 0$.

The 3D plot of Figure 2.2-2 confirms the 2D plot of Figure 2.2-1. Also, the variation of the solution in x reflects the selected value $d = 0.05$. The effect of variations in the diffusivity can easily be studied and this is left as an exercise.

2.1.5 Numerical, graphical output, Robin BCs

If Robin BCs,[2]

$$\frac{\partial u(x = x_l, t)}{\partial x} = k_b u(x = x_l, t)$$

$$\frac{\partial u(x = x_u, t)}{\partial x} = -k_b u(x = x_u, t) \tag{2.1-7,8}$$

are considered, the programming in `pde1a` changes.

---

[2]Robin BCs include the dependent variable, $u(x = x_l, x_u, t)$, and its derivative $\dfrac{\partial u(x = x_l, x_u, t)}{\partial x}$.

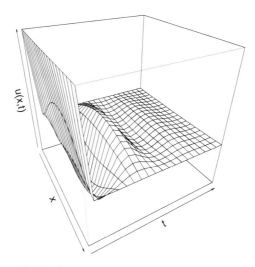

Figure 2.2-2 Numerical solution $u(x,t)$ from eq. (2.1-1), $d = 0.05$, BCs (2.1-5,6), persp.

```
#
# ux
  ux=dss004(xl,xu,nx,u);
#
# BCs
  ux[1]=   10*u[1];
  ux[nx]=-10*u[nx];
#
# uxx
  nl=2;nu=2;
  uxx=dss044(xl,xu,nx,u,ux,nl,nu);
```
Listing 2.5 Programming for Robin BCs (eqs. (2.1-7,8)).

The numerical output is in Table 2.5.

```
[1] 21

[1] 22
```

t	x	u(x,t)
0.00e+00	0.00e+00	1.0000
0.00e+00	2.50e-01	1.0000
0.00e+00	5.00e-01	1.0000
0.00e+00	7.50e-01	1.0000
0.00e+00	1.00e+00	1.0000

t	x	u(x,t)
2.50e+00	0.00e+00	-0.0516

```
2.50e+00    2.50e-01       -0.1869
2.50e+00    5.00e-01       -0.2620
2.50e+00    7.50e-01       -0.1869
2.50e+00    1.00e+00       -0.0516


    t              x          u(x,t)
5.00e+00    0.00e+00       -0.0005
5.00e+00    2.50e-01       -0.0029
5.00e+00    5.00e-01       -0.0049
5.00e+00    7.50e-01       -0.0029
5.00e+00    1.00e+00       -0.0005


    t              x          u(x,t)
7.50e+00    0.00e+00        0.0082
7.50e+00    2.50e-01        0.0255
7.50e+00    5.00e-01        0.0322
7.50e+00    7.50e-01        0.0255
7.50e+00    1.00e+00        0.0082


    t              x          u(x,t)
1.00e+01    0.00e+00       -0.0044
1.00e+01    2.50e-01       -0.0137
1.00e+01    5.00e-01       -0.0173
1.00e+01    7.50e-01       -0.0137
1.00e+01    1.00e+00       -0.0044


ncall = 2078
```

Table 2.5   Abbreviated output from Listings 2.1 and 2.5, $d = 0.05$, BCs (eq. (2.1-7,8)).

The solution from BCs (2.1-7,8) indicates a maximum or minimum at the midpoint $x = (x_u - x_l)/2 = 0.5$ (depending on the value of t). The computational effort is modest at ncall = 2078.

This 2D plot of $u(x,t)$ against x demonstrates the variation with x. Note that BCs (2.1-7,8) introduce a variation in $u(x,t)$ at the boundaries $x = x_l, x_u$ from the initial values $u(x = x_l, x_u, t = 0) = 1$ to varying boundary values which for large t approach $u(x = x_l, x_u, t \to \infty) = 0$.

The 3D plot of Figure 2.3-2 is consistent with the 2D plot of Figure 2.3-1. Also, the variation of the solution in x reflects the constant k_b in the Robin BCs (2.1-7,8) programmed in pde1a. This constant can be considered as a ratio

$$k_b = \frac{\text{boundary mass transfer coefficient}}{\text{diffusivity } d} = 10$$

and was selected (as $k_b = 10$) to give a smooth transition from the IC to the BCs at the boundaries. The effect of variations in this constant can easily be studied and this is left as an exercise.

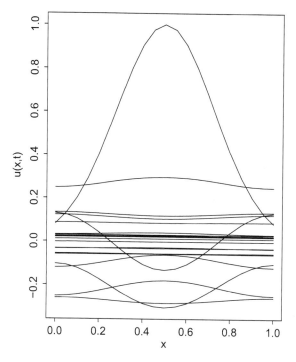

Figure 2.3-1   Numerical solution $u(x,t)$ from eq. (2.1-1), $d\ =\ 0.05$, BCs (2.1-7,8), `matplot`.

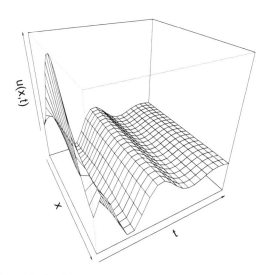

Figure 2.3-2   Numerical solution $u(x,t)$ from eq. (2.1-1), $d\ =\ 0.05$, BCs (2.1-7,8), `persp`.

This case demonstrates the ease of using Robin BCs. Extensions can include nonlinear Robin BCs such as

$$\frac{\partial u(x = x_l, t)}{\partial x} = k_b(u(x = x_l, t)^4 - u_a^4)$$

$$\frac{\partial u(x = x_u, t)}{\partial x} = -k_b(u(x = x_u, t)^4 - u_a^4)$$

where u_a is an ambient value.[3]

As a concluding case, the uniform IC programmed in Listing 2.1 is replaced with a Gaussian IC.

2.1.6 Numerical, graphical output, Gaussian IC

The IC/history vector in Listing 2.1 is a Gaussian function centered at $x = 0.5$ (with the homogeneous Neumann BCs of Listing 2.2, $d = 0.05$).

```
#
# IC vector
  u0=rep(0,nx);
  for(i in 1:nx){
    u0[i]=exp(-10*(x[i]-0.5)^2);
  }
```

Listing 2.6   Programming for Gaussian IC.

The numerical output is in Table 2.6.

```
[1] 21

[1] 22
```

t	x	u(x,t)
0.00e+00	0.00e+00	0.0821
0.00e+00	2.50e-01	0.5353
0.00e+00	5.00e-01	1.0000
0.00e+00	7.50e-01	0.5353
0.00e+00	1.00e+00	0.0821

t	x	u(x,t)
2.50e+00	0.00e+00	-0.2519
2.50e+00	2.50e-01	-0.2163
2.50e+00	5.00e-01	-0.1806
2.50e+00	7.50e-01	-0.2163
2.50e+00	1.00e+00	-0.2519

t	x	u(x,t)
5.00e+00	0.00e+00	0.0879
5.00e+00	2.50e-01	0.0865

---

[3]Physically, these nonlinear BCs can be considered to model radiation at the boundaries.

t	x	u(x,t)
5.00e+00	5.00e-01	0.0851
5.00e+00	7.50e-01	0.0865
5.00e+00	1.00e+00	0.0879
t	x	u(x,t)
7.50e+00	0.00e+00	-0.0318
7.50e+00	2.50e-01	-0.0324
7.50e+00	5.00e-01	-0.0331
7.50e+00	7.50e-01	-0.0324
7.50e+00	1.00e+00	-0.0318
t	x	u(x,t)
1.00e+01	0.00e+00	0.0111
1.00e+01	2.50e-01	0.0111
1.00e+01	5.00e-01	0.0110
1.00e+01	7.50e-01	0.0111
1.00e+01	1.00e+00	0.0111

```
ncall = 1624
```

Table 2.6   Abbreviated output from Listings 2.1 and 2.6, $d = 0.05$, Gaussian IC

The numerical solution, $u(x, t)$, starts with the Gaussian at $t = 0$, and approaches a uniform value of 0.0111 as reflected in Table 2.6 and the graphical output, Figures 2.4.

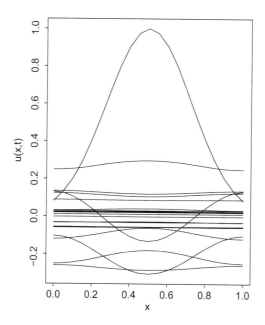

Figure 2.4-1   Numerical solution $u(x, t)$ from eq. (2.1-1), $d = 0.05$, Gaussian IC, matplot.

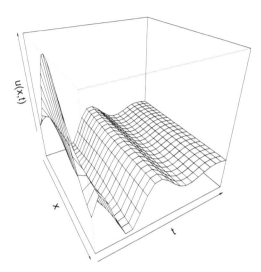

Figure 2.4-2 Numerical solution $u(x,t)$ from eq. (2.1-1), $d = 0.05$, Gaussian IC, `persp`.

Since there is no diffusion through the boundaries (homogeneous Neumann BCs), the decrease in $u(x,t)$ is from the lag term in eq. (2.1-1). This can be confirmed by using $a = 0$, and is left an an exercise (the solution approaches $u(x, t \to \infty) = 0.5463$ reflecting conservation of mass).

2.2   Convection diffusion reaction delayed PDE

Eq. (1.2-1) is now extended to a convection diffusion reaction delayed PDE (CDR DPDE).

$$\frac{\partial u(x,t)}{\partial t} = d\frac{\partial^2 u(x,t)}{\partial x^2} - v\frac{\partial u(x,t)}{\partial x} - au(x,t-1) - bu(x,t) \qquad (2.2\text{-}1)$$

with Neumann BCs

$$\frac{\partial u(x = x_l, t)}{\partial x} = f(t); \quad \frac{\partial u(x = x_u, t)}{\partial x} = 0 \qquad (2.2\text{-}2,3)$$

and IC

$$u(x, t = 0) = u_0(x) \qquad (2.2\text{-}4)$$

The left BC eq. (2.2-2) can now be a function of t, $f(t)$, to reflect an incoming convective flux at $x = x_l$. Also, the following discussion of eqs. (2.2) is for a Gaussian IC, $u_0(x) = e^{-10x^2}$.

A main program for eqs. (2.2) is similar to the main program for eqs. (2.1), so only the differences are included in Listing 2.7.

2.2.1   Main program

```
#
# Basic DPDE with convection
#
# Delete previous workspaces
  rm(list=ls(all=TRUE))
#
# Access DODE integrator
  library(deSolve)
#
# Access functions for numerical solution
  setwd("f:/dpde/chap2");
  source("pde1b.R");
  source("dss004.R");
  source("dss044.R");
#
# Model parameters
  tau=1;
  a=0;
  b=0;
  d=0.05;
  v=0.5;
          .
          .
          .

#
# IC vector
  u0=rep(0,nx);
  for(i in 1:nx){
    u0[i]=exp(-10*(x[i]-0.5)^2);
  }
  ncall=0;
          .
          .
          .

#
# Integration of delay MOL/ODEs
  uout=dede(y=u0,times=times,func=pde1b);
  nrow(uout);
  ncol(uout);
          .
          .
          .
```

Listing 2.7   Programming for eqs. (2.2).

We can note the following details about Listing 2.7.

- The ODE/MOL routine is pde1b (discussed next).

- The model parameters include the velocity v in the convection term of eq. (2.2-1).

- IC (2.2-4) is a Gaussian function centered at $r = 0.5$.

Otherwise, the main program is the same as in Listing 2.1.

2.2.2   DODE routine

An abbreviated listing of the ODE/MOL routine pde1b follows.

```
pde1b=function(t,u,parm,lag) {
        .
        .
        .
#
#  DPDE
  ut=rep(0,nx);
  for(i in 1:nx){
    ut[i]=d*uxx[i]-v*ux[i]-a*ulag[i]-b*u[i];
  }
        .
        .
        .
```

Listing 2.8   Abbreviated MOL/ODE routine for eqs. (2.2).

The differences between Listings 2.2 and 2.8 are: (1) the name of the routine is pde1b rather than pde1a, and (2) the addition of the convection term $-v\frac{\partial u}{\partial x}$. The first derivative in x is computed by dss004 (as in Listing 2.2).

2.2.3   Numerical, graphical output

A series of special cases for the main program of Listing 2.7 and the subordinate routine pde1b of Listing 2.8 follows. The abbreviated numerical output for the parameters in Listing 2.7 is given in Table 2.7.

```
[1]  21

[1]  22

          t              x          u(x,t)
   0.00e+00      0.00e+00          0.0821
   0.00e+00      2.50e-01          0.5353
   0.00e+00      5.00e-01          1.0000
   0.00e+00      7.50e-01          0.5353
   0.00e+00      1.00e+00          0.0821
         .              .               .
         .              .               .
         .              .               .
```

```
Output for t=2.5,5,7.5 removed
              .                    .
              .                    .
              .                    .
        t             x           u(x,t)
   1.00e+01      0.00e+00         0.2301
   1.00e+01      2.50e-01         0.2301
   1.00e+01      5.00e-01         0.2301
   1.00e+01      7.50e-01         0.2301
   1.00e+01      1.00e+00         0.2301

ncall = 1304
```

Table 2.7 Abbreviated output from Listings 2.1 and 2.7

We can note the following details about this output.

- 21 t output points as the first dimension of the solution matrix `uout` from `dede` as programmed in the main program of Listings 2.1 and 2.7.

- The solution matrix `uout` returned by `dede` has 22 elements as a second dimension. The first element is the value of t. Elements 2–22 are $u(x,t)$ from eq. (2.2-1) (for each of the 21 output points).

- The solution is displayed for $t = 0, 2.5, ..., 10$ and $x = 0, 0.25, ..., 1$ as programmed in Listing 2.1.

- The solution is greater than in Table 2.6 since the delay and current depletion terms are not included ($a = b = 0$) (these terms are negative and therefore move the solution to smaller values, e.g., $u(x, t = 10) = 0.0101$, Table 2.6, $a = 1, b = 0$ or $u(x, t = 10) = 0.2301$, Table 2.7, $a = 0, b = 0$).

- The computational effort is modest, `ncall` $= 1304$, so that `dede` computed a solution to eqs. (2.2) efficiently and accurately.

The graphical output is in Figures 2.5.

The solution is complicated but the movement of the IC Gaussian left to right due to the convective term in eq. (2.2-1) is clear. Eqs. (2.2) are a hyperbolic–parabolic (convective–diffusive) DPDE system.[4]

---

[4]Hyperbolic systems can be difficult to integrate numerically because: (1) they can propagate steep moving fronts and discontinuties, and (2) the solutions can display numerical artifacts, particularly at the exit (outflow) boundary ($x = x_u = 1$) (typically numerical oscillations). In the present case, these numerical problems do not develop due to the smoothing of the diffusion (from the second-order spatial term in eq. (2.2-1)). Stated quantitatively, the Peclet number is small, $P_e = \dfrac{Lv}{d} = \dfrac{(1)(0.5)}{0.05} = 10$ with $L = x_u - x_l = 1$. For larger P_e, for example, > 100, special approximations of the convective term are required, such as flux limiters, to avoid numerical problems.

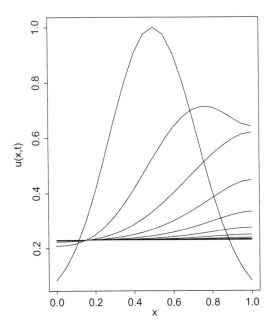

Figure 2.5-1 Numerical solution $u(x,t)$ from eq. (2.2-1), $a = b = 0, v = 0.5$, Gaussian IC, `matplot`.

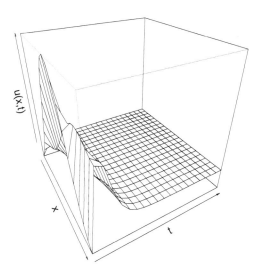

Figure 2.5-2 Numerical solution $u(x,t)$ from eq. (2.2-1), $a = b = 0, v = 0.5$, Gaussian IC, `persp`.

In summary, the steady state solution $u(x, t = 10) = 0.2301$ (constant in x, Table 2.7) follows from eq. (2.2-1) since the derivatives in x are zero (derivative of a constant) and the two additional RHS terms are zero ($a = b = 0$) so that the LHS is $\dfrac{\partial u}{\partial t} = 0$ (steady state). With the homogeneous (no flux) BCs (2.2-2,3), the area under the IC Gaussian (eq. (2.2-4)) is redistributed to a uniform value of 0.2301 (Figure 2.5-1).

As the next case, current (concurrent) depletion is added to the RHS of eq. (2.2-1) with $b = 1$ (the only change in the main program of Listing 2.1 and pde1b unchanged). The abbreviated output follows in Table 2.8.

```
[1] 21

[1] 22


      t              x          u(x,t)
0.00e+00      0.00e+00         0.0821
0.00e+00      2.50e-01         0.5353
0.00e+00      5.00e-01         1.0000
0.00e+00      7.50e-01         0.5353
0.00e+00      1.00e+00         0.0821

       .             .             .
       .             .             .
       .             .             .
Output for t=2.5,5,7.5 removed
       .             .             .
       .             .             .
       .             .             .

      t              x          u(x,t)
1.00e+01      0.00e+00         0.0000
1.00e+01      2.50e-01         0.0000
1.00e+01      5.00e-01         0.0000
1.00e+01      7.50e-01         0.0000
1.00e+01      1.00e+00         0.0000

ncall = 1267
```

Table 2.8   Abbreviated output from Listings 2.1 and 2.7, $a = 0, b = 1$.

The steady state solution is now $u(x, t = 10) = 0.0000$ indicating the effect of the depletion with $b = 1$ (and homogeneous Neumann BCs). An important feature of this solution is that as $\dfrac{\partial u}{\partial t}$ is reduced by $-bu(x, t)$ in eq. (2.2-1), $u(x, t) \geq 0$ (monotonically).

These features of the solution are also indicated in the graphical output (Figures 2.6).

As a concluding case, both depletion terms in eq. (2.2-1) are included with $a = b = 1$. The output follows in Table 2.9.

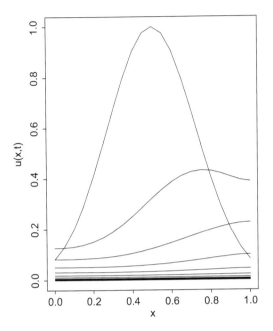

Figure 2.6-1 Numerical solution $u(x,t)$ from eq. (2.2-1), $a = 0, b = 1, v = 0.5$, Gaussian IV, `matplot`.

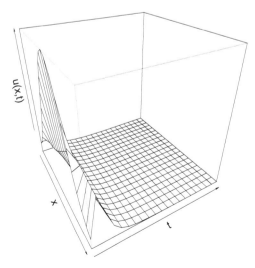

Figure 2.6-2 Numerical solution $u(x,t)$ from eq. (2.2-1), $a = 0, b = 1, v = 0.5$, Gaussian IC, `persp`.

[1] 21

[1] 22

t	x	u(x,t)
0.00e+00	0.00e+00	0.0821
0.00e+00	2.50e-01	0.5353
0.00e+00	5.00e-01	1.0000
0.00e+00	7.50e-01	0.5353
0.00e+00	1.00e+00	0.0821

t	x	u(x,t)
2.50e+00	0.00e+00	0.0015
2.50e+00	2.50e-01	0.0110
2.50e+00	5.00e-01	0.0500
2.50e+00	7.50e-01	0.0833
2.50e+00	1.00e+00	0.0737

t	x	u(x,t)
5.00e+00	0.00e+00	-0.0104
5.00e+00	2.50e-01	-0.0101
5.00e+00	5.00e-01	-0.0084
5.00e+00	7.50e-01	-0.0066
5.00e+00	1.00e+00	-0.0071

t	x	u(x,t)
7.50e+00	0.00e+00	0.0009
7.50e+00	2.50e-01	0.0009
7.50e+00	5.00e-01	0.0009
7.50e+00	7.50e-01	0.0007
7.50e+00	1.00e+00	0.0004

t	x	u(x,t)
1.00e+01	0.00e+00	0.0004
1.00e+01	2.50e-01	0.0004
1.00e+01	5.00e-01	0.0004
1.00e+01	7.50e-01	0.0004
1.00e+01	1.00e+00	0.0003

ncall = 1955

Table 2.9 Abbreviated output from Listings 2.1 and 2.7, $a = 1, b = 1$.

The steady state solution is again $u(x, t \to \infty) = 0$ (the small values at $t = 10$ become zero for larger t). However, $u(x,t)$ oscillates, including negative values, as it approaches zero (see the $t = 5$ output). This oscillation occurs because past values of $u(x,t)$ are used in the delay term $-au(x, t - \tau)$ that are larger than the current values of $u(x,t)$.

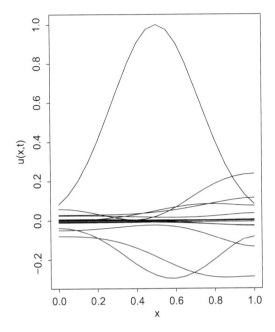

Figure 2.7-1   Numerical solution $u(x, t)$ from eq. (2.2-1), $a = b = 1, v = 0.5$, Gaussian IV, `matplot`.

These features of the solution are also indicated in the graphical output (Figures 2.7).

This concludes the discussion of special case solutions of eqs. (2.2). In all of these cases, the convection from the term $-v\dfrac{\partial u}{\partial x}$ is clear with the movement of the Gaussian IC centered at $x = 0.5$ from left to right (and no complications at the outflow boundary $x = x_u = 1$).

2.3   Summary and conclusions

Additional cases could be considered. For example, the effect of the delay (lag) in eqs. (2.1-1), (2.2-1) could be studied by varying τ, starting with $\tau = 0$ (no delay). The oscillation of the DPDEs would be of particular interest. The study of τ is left as an exercise.

In summary, the model of eqs. (2.1) and (2.2) demonstrate the MOL programming of a DPDE system. For eqs. (2.1), the three basic forms of BCs, Dirichlet, Neumann, Robin, are examined as three cases. Also, the effect of

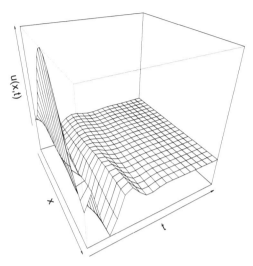

Figure 2.7-2   Numerical solution $u(x, t)$ from eq. (2.2-1), $a = b = 1, v = 0.5$, Gaussian IC, `persp`.

diffusion is examined by varying d. For eqs. (2.2), the oscillatory effect of the delay term $-au(x, t - \tau)$ is demonstrated.

The investigation of the effect of the RHS terms of eq. (2.1-1), (2.2-1) can be degeneralized by computing and displaying the terms individually. This procedure is discussed in [2, Chapter 4].

The DPDE methodology that is discussed and explained through the basic DPDEs of eqs. (2.1), (2.2) is now applied in subsequent chapters to a series of DPDEs with emphasis on BMSE applications.

References

[1] Schiesser, W.E. (2016), *Method of Lines PDE Analysis in Biomedical Science and Engineering*, Wiley, Hoboken, NJ.

[2] Schiesser, W.E. (2019), *Numerical PDE Analysis of Retinal Neovascularization*, Elsevier, Cambridge, MA.

[3] Soetaert, K., J. Cash, and F. Mazzia (2012), *Solving Differential Equations in R*, Springer-Verlag, Heidelberg, Germany.

3

Hepatitis B Virus DPDE Model

Introduction

The DPDE methodology discussed in Chapter 2 is now applied to a 4×4 (4 DPDEs in 4 unknowns) system that models the spatiotemporal variation of a hepatitis B virus (HBV). The discussion starts with two abbreviated quotations from Xu and Ma [3].

[3, p. 499] (references not included).

(1) Persistent infection with HBV is a major health problem worldwide as it can lead to cirrhosis and primary hepatocellular carcinoma. Chronic HBV infection is often the result of exposure early in life, leading to viral persistence in the absence of strong antibodies or cellular immune responses. Therapy of HBV carriers generally aims to either inhibit viral replication or enhance immunological responses against viruses, or both.

[3, p. 500] (references not included)

(2) To account for the time between viral entry into a target cell and the production of new virus particles, models that include delays have been introduced.

The following model is an SIRV[1] extension of the HBV model reported in [3], expressed in 1D spherical coordinates to represent a domain (differential volume) of an HBV infection.

3.1 DPDE model

The 4×4 DPDE system in coordinate-free format is

$$\frac{\partial S}{\partial t} = D_s \nabla^2 S + S_p - c_1 \frac{SV}{1 + c_2 V} - c_3 S \tag{3.1-1}$$

---

[1]SIRV $\rightarrow$ Susceptible Infected Recovered Virus (or Vaccinated).

$$\frac{\partial I}{\partial t} = D_i \nabla^2 I + c_1 \frac{S_{t-\tau} V_{t-\tau}}{1 + c_2 V_{t-\tau}} - c_4 I \tag{3.1-2}$$

$$\frac{\partial R}{\partial t} = D_r \nabla^2 R + c_5 I - c_6 R \tag{3.1-3}$$

$$\frac{\partial V}{\partial t} = D_v \nabla^2 V + c_7 I - c_8 V \tag{3.1-4}$$

The variables and parameters of eqs. (3.1) are listed and briefly explained in Table 3.1.

Variable, parameter	Description
$S(r,t)$	susceptible cells
$I(r,t)$	infected cells
$R(r,t)$	recovered cells
$V(r,t)$	free virus (HBV) concentration
∇	gradient of a scalar
$\nabla \cdot$	divergence of a vector
$\nabla \cdot \nabla = \nabla^2$	Laplacian differential operator
Sp	rate of production of susceptible cells
D_s, D_i, D_r, D_v	diffusivities
τ	delay in infection from HBV
c_1 to c_8	rate constants

Table 3.1   Variables and parameters of HBV model

If the spatial domain is expressed in 1D spherical coordinates (the two angular components in (r, ϕ, θ) are deleted), eqs. (3.1) are

$$\frac{\partial S(r,t)}{\partial t} = D_s \left(\frac{\partial^2 S(r,t)}{\partial r^2} + \frac{2}{r} \frac{\partial S(r,t)}{\partial r} \right) + S_p - c_1 \frac{S(r,t)V(r,t)}{1 + c_2 V(r,t)} - c_3 S(r,t) \tag{3.2-1}$$

A brief explanation of the terms in eq. (3.2-1) follows.

- $\dfrac{\partial S(r,t)}{\partial t}$: Net accumulation or depletion (for this term positive or negative) of susceptible cells in an incremental volume $4\pi r^2 \Delta r$.

  The incremental volume is the difference in the volumes at $r + \Delta r$ and r:

$$(4/3)\pi (r + \Delta r)^3 - (4/3)\pi r^3 = (4/3)\pi (r^2 + 2r\Delta r + \Delta r^2)(r + \Delta r) - (4/3)\pi r^3 =$$

$$(4/3)\pi (r^3 + 2r^2 \Delta r + r\Delta r^2 + r^2 \Delta r + 2r\Delta r^2 + \Delta r^3) - (4/3)\pi r^3 =$$

$$(4/3)\pi (3r^2 \Delta r + 3r\Delta r^2 + \Delta r^3)$$

which for small Δr is

$$4\pi r^2 \Delta r$$

- $D_s \left(\dfrac{\partial^2 S(r,t)}{\partial r^2} + \dfrac{2}{r} \dfrac{\partial S(r,t)}{\partial r} \right)$: Net diffusion of susceptible cells into or out of the incremental volume $4\pi r^2 \Delta r$.

This term follows from

$$\frac{4\pi r^2 D_s \dfrac{\partial S(r,t)}{\partial r}\big|_{r+\Delta r} - 4\pi r^2 D_s \dfrac{\partial S(r,t)}{\partial r}\big|_r}{4\pi r^2 \Delta r}$$

$$= D_s \frac{1}{r^2} \left(\frac{r^2 \dfrac{\partial S(r,t)}{\partial r}\big|_{r+\Delta r} - r^2 \dfrac{\partial S(r,t)}{\partial r}\big|_r}{\Delta r} \right)$$

which for $\Delta r \to 0$ is

$$D_s \frac{1}{r^2} \frac{\partial (r^2 \dfrac{\partial S(r,t)}{\partial r})}{\partial r} = D_s \left(\frac{\partial^2 S(r,t)}{\partial r^2} + \frac{2}{r} \frac{\partial S(r,t)}{\partial r} \right)$$

- $+S_p$: Volumetric rate of production of susceptible cells.

- $-c_1 \dfrac{S(r,t)V(r,t)}{1 + c_2 V(r,t)}$: Volumetric rate of reduction of susceptible cells from the combined effect of $S(r,t)$ and $V(r,t)$ according to Michaelis–Menten kinetics, with $c_1 > 0$.

- $-c_3 S(r,t)$: Volumetric rate of change of susceptible cells from natural causes, for example, migrations, births, and deaths.

The DPDE for the infected cell density, $I(r,t)$, is

$$\frac{\partial I(r,t)}{\partial t} = D_i \left(\frac{\partial^2 I(r,t)}{\partial r^2} + \frac{2}{r} \frac{\partial I(r,t)}{\partial r} \right) + c_1 \frac{S(r,t-\tau)V(r,t-\tau)}{1 + c_2 V(r,t-\tau)} - c_4 I(r,t)$$

$$(3.2\text{-}2)$$

A brief explanation of the terms in eq. (3.2-2) follows.

- $\dfrac{\partial I(r,t)}{\partial t}$: Net accumulation or depletion (for this term positive or negative) of infected cells in the differential volume $4\pi r^2 dr$.

- $D_i \left(\dfrac{\partial^2 I(r,t)}{\partial r^2} + \dfrac{2}{r} \dfrac{\partial I(r,t)}{\partial r} \right)$: Net diffusion of infected cells into or out of the differential volume $4\pi r^2 dr$.

- $+c_1 \dfrac{S(r,t-\tau)V(r,t-\tau)}{1 + c_2 V(r,t-\tau)}$: Net volumetric rate of generation of infected cells from the combined effect of $S(r,t)$ and $V(r,t)$, according to Michaelis–Menten kinetics, with $c_1 > 0$. $S(r,t), V(r,t)$ are delayed (lagged) to reflect the time required for the infected cells to form from the susceptible cells and HBV.

- $-c_4I(r,t)$: Volumetric rate of change of infected cells from natural causes, for example, migrations, births, and deaths.

The DPDE for the recovered cell density, $R(r,t)$, is

$$\frac{\partial R(r,t)}{\partial t} = D_r \left(\frac{\partial^2 R(r,t)}{\partial r^2} + \frac{2}{r} \frac{\partial R(r,t)}{\partial r} \right) + c_5 I(r,t) - c_6 R(r,t) \qquad (3.2\text{-}3)$$

A brief explanation of the terms in eq. (3.2-3) follows.

- $\dfrac{\partial R(r,t)}{\partial t}$: Net accumulation or depletion (for this term positive or negative) of recovered cells in the differential volume $4\pi r^2 dr$.

- $D_r \left(\dfrac{\partial^2 R(r,t)}{\partial r^2} + \dfrac{2}{r} \dfrac{\partial R(r,t)}{\partial r} \right)$: Volumetric diffusion of recovered cells into or out of the differential volume $4\pi r^2 dr$.

- $+c_5 I(r,t)$: Volumetric rate of increase of recovered cells from infected cells.

- $-c_6 R(r,t)$: Volumetric rate of change of recovered cells from natural causes, for example, migrations, births, and deaths.

The DPDE for the HBV concentration, $V(r,t)$, is

$$\frac{\partial V(r,t)}{\partial t} = D_v \left(\frac{\partial^2 V(r,t)}{\partial r^2} + \frac{2}{r} \frac{\partial V(r,t)}{\partial r} \right) + c_7 I(r,t) + c_8 V(r,t) \qquad (3.2\text{-}4)$$

A brief explanation of the terms in eq. (3.2-4) follows.

- $\dfrac{\partial V(r,t)}{\partial t}$: Net accumulation or depletion (for this term positive or negative) of HBV in the differential volume $4\pi r^2 dr$.

- $D_v \left(\dfrac{\partial^2 V(r,t)}{\partial r^2} + \dfrac{2}{r} \dfrac{\partial V(r,t)}{\partial r} \right)$: Volumetric diffusion of HBV into or out of the differential volume $4\pi r^2 dr$.

- $+c_7 I(r,t)$: Volumetric rate of increase of HBV from infected cells.

- $+c_8 V(r,t)$: Volumetric rate of change of HBV cells from natural causes, for example, migrations, births ($c_8 > 0$), and deaths ($c_8 < 0$).

Eqs. (3.2), are first order in t and therefore each requires one initial condition (IC).

$$S(r, t = 0) = S_0(r) \qquad (3.3\text{-}1)$$
$$I(r, t = 0) = I_0(r) \qquad (3.3\text{-}2)$$
$$R(r, t = 0) = R_0(r) \qquad (3.3\text{-}3)$$
$$V(r, t = 0) = V_0(r) \qquad (3.3\text{-}4)$$

As a requirement of the delay term in eq. (3.2-2), $\dfrac{S(r, t-\tau)V(r, t-\tau)}{1 + c_3 V(r, t-\tau)}$, these IC functions also provide history vectors (as implemented in the DODE routine pdela considered subsequently). Therefore, eqs. (3.3) apply over the interval $t \in [-\tau, 0]$.

Eqs. (3.2) are second order in r so that each requires two boundary conditions (BCs).

$$\frac{\partial S(r=0,t)}{\partial r} = \frac{\partial S(r=r_u,t)}{\partial r} = 0 \qquad\qquad (3.4\text{-}1,2)$$

$$\frac{\partial I(r=0,t)}{\partial r} = \frac{\partial I(r=r_u,t)}{\partial r} = 0 \qquad\qquad (3.4\text{-}3,4)$$

$$\frac{\partial R(r=0,t)}{\partial r} = \frac{\partial R(r=r_u,t)}{\partial r} = 0 \qquad\qquad (3.4\text{-}5,6)$$

$$\frac{\partial V(r=0,t)}{\partial r} = \frac{\partial V(r=r_u,t)}{\partial r} = 0 \qquad\qquad (3.4\text{-}7,8)$$

The BCs at $r = 0$ reflect symmetry. The homogeneous Neumann BCs at $r = r_u$ specify no flux or diffusion through the outer boundary.

Eqs. (3.2), (3.3), and (3.4) constitute the 4×4 DPDE model. The MOL solution of this system is considered next, starting with a main program.

3.1.1 Main program

The main program for eqs. (3.2), (3.3) and (3.4) follows.

```
#
# HBV model
#
# Delete previous workspaces
  rm(list=ls(all=TRUE))
#
# Access DODE integrator
  library(deSolve)
#
# Access functions for numerical solution
  setwd("f:/dpde/chap3");
  source("pdela.R");
  source("dss004.R");
  source("dss044.R");
#
# Model parameters
  ncase=1;
  tau=5;
```

```
#
#   S(r,t)
    Ds=1.0e-08;Sp=1.0e+01;c1=1.0e-04;
                c2=1.0e-04;c3=5.0e-03;
#
#   I(r,t)
    Di=1.0e-08;c4=5.0e-02;
#
#   R(r,t)
    Dr=1.0e-08;c5=1.0e-02;c6=5.0e-03;
#
#   V(r,t)
    Dv=1.0e-08;c7=1.0e-03;
    if(ncase==1){c8= 5.0e-03;}
    if(ncase==2){c8=-5.0e-03;}
#
# Functions
  fS0=function(r) 3.0e+03;
  fI0=function(r) 1.0e+03;
  fR0=function(r) 5.0e+02;
  fV0=function(r) 1.0e+03*exp(-25*r^2);
#
# Spatial grid
  nr=41;rl=0;ru=1;
  r=seq(from=rl,to=ru,by=(ru-rl)/(nr-1));
#
# Temporal grid
  nout=21;t0=0;tf=200;
  times=seq(from=t0,to=tf,by=(tf-t0)/(nout-1));
#
# IC vector
  u0=rep(0,4*nr);
  for(i in 1:nr){
    u0[i]     =fS0(r[i]);
    u0[i+nr]  =fI0(r[i]);
    u0[i+2*nr]=fR0(r[i]);
    u0[i+3*nr]=fV0(r[i]);
  }
  ncall=0;
#
# Integration of DPDEs
  uout=dede(y=u0,times=times,func=pde1a);
  nrow(uout);
  ncol(uout);
#
# Arrays/vectors for ODE solutions
  Sp=matrix(0,nrow=nr,ncol=nout);
  Ip=matrix(0,nrow=nr,ncol=nout);
  Rp=matrix(0,nrow=nr,ncol=nout);
```

```
  Vp=matrix(0,nrow=nr,ncol=nout);
  tp=rep(0,nout);
  for(it in 1:nout){
    tp[it]=uout[it,1];
  for( i in 1:nr){
    Sp[i,it]=uout[it,i+1];
    Ip[i,it]=uout[it,i+1+nr];
    Rp[i,it]=uout[it,i+1+2*nr];
    Vp[i,it]=uout[it,i+1+3*nr];
  }
  }
#
# Display numerical solution
  iv=seq(from=1,to=nout,by=5);
  for(it in iv){
    cat(sprintf("\n                t"));
    cat(sprintf("\n%12.1f\n",tp[it]));
  iv=seq(from=1,to=nr,by=10);
  for(i in iv){
    cat(sprintf("\n                r"));
    cat(sprintf("\n%12.2f\n",r[i]));
    cat(sprintf("       S(r,t)        I(r,t)"));
    cat(sprintf("\n       R(r,t)        V(r,t)"));
    cat(sprintf("\n%12.2e%12.2e",  Sp[i,it],Ip[i,it]));
    cat(sprintf("\n%12.2e%12.2e\n",Rp[i,it],Vp[i,it]));
  }
  }
#
# Display calls to pde1a
  cat(sprintf("\n ncall = %4d\n",ncall));
#
# Plot DPDE solutions
  par(mfrow=c(1,1));
#
# 2D
  matplot(r,Sp,type="l",xlab="r",ylab="S(r,t)",
          lty=1,main="",lwd=2,col="black");
  matplot(r,Ip,type="l",xlab="r",ylab="I(r,t)",
          lty=1,main="",lwd=2,col="black");
  matplot(r,Rp,type="l",xlab="r",ylab="R(r,t)",
          lty=1,main="",lwd=2,col="black");
  matplot(r,Vp,type="l",xlab="r",ylab="V(r,t)",
          lty=1,main="",lwd=2,col="black");
#
# 3D
  persp(r,tp,Sp,theta=60,phi=30,
        xlim=c(rl,ru),ylim=c(t0,tf),xlab="r",
        ylab="t",zlab="S(r,t)");
  persp(r,tp,Ip,theta=60,phi=30,
```

```
      xlim=c(rl,ru),ylim=c(t0,tf),xlab="r",
      ylab="t",zlab="I(r,t)");
persp(r,tp,Rp,theta=60,phi=30,
      xlim=c(rl,ru),ylim=c(t0,tf),xlab="r",
      ylab="t",zlab="R(r,t)");
persp(r,tp,Vp,theta=60,phi=30,
      xlim=c(rl,ru),ylim=c(t0,tf),xlab="r",
      ylab="t",zlab="V(r,t)");
```

Listing 3.1   Main program for eqs. (3.2), (3.3), (3.4).

The following discussion of Listing 3.1 partly repeats the discussion of Listings 1.1 and 2.1 in Chapters 1 and 2, respectively, but is given so this discussion of the HBV model is self-contained.

- Previous workspaces are deleted.

```
#
# HBV model
#
# Delete previous workspaces
  rm(list=ls(all=TRUE))
```

- The R ODE integrator library deSolve is accessed. Then the directory with the files for the solution of eqs. (3.2) is designated. Note that setwd (set working directory) uses / rather than the usual \.

```
#
# Access DODE integrator
  library(deSolve)
#
# Access functions for numerical solution
  setwd("f:/dpde/chap3");
  source("pde1a.R");
  source("dss004.R");
  source("dss044.R");
```

pde1a.R is the routine for eqs. (3.2) (discussed subsequently) based on the MOL, a general algorithm for PDEs [1]. dss004, dss044 are library routines for the calculation of first and second spatial derivatives. These routines are listed in Appendix A1 with additional explanation.

- The parameters of eqs. (3.2) are specified (ncase is discussed next).

```
#
# Model parameters
  ncase=1;
  tau=5;
#
#   S(r,t)
    Ds=1.0e-08;Sp=1.0e+01;c1=1.0e-04;
            c2=1.0e-04;c3=5.0e-03;
```

```
#
#      I(r,t)
       Di=1.0e-08;c4=5.0e-02;
#
#      R(r,t)
       Dr=1.0e-08;c5=1.0e-02;c6=5.0e-03;
#
#      V(r,t)
       Dv=1.0e-08;c7=1.0e-03;
       if(ncase==1){c8= 5.0e-03;}
       if(ncase==2){c8=-5.0e-03;}
```

c_1, c_2 were selected so that the nonlinear term $c_1 \dfrac{SV}{1 + c_2 V}$ has an effect comparable to the other terms in eqs. (3.1-1,2). c_3, c_4, c_6, c_8 were selected to give significant increase or decay of the solutions over $0 \le t \le 200$. c_5, c_7 were selected so that the transition of the infected cells to the recovered cells and the HBV is significant.

- The IC functions in eqs. (3.3) are defined.

```
#
# Functions
  fS0=function(r)  3.0e+03;
  fI0=function(r)  1.0e+03;
  fR0=function(r)  5.0e+02;
  fV0=function(r)  1.0e+03*exp(-25*r^2);
```

IC (3.3-4) is a Gaussian function, $V_0(r) = 10^3 e^{-25r^2}$.

- A spatial grid is defined for nr=41 points in r, with $r_l = 0, r_u = 1$ in BCs (3.4), so $r = 0, 0.025, ..., 1$.

```
#
# Spatial grid
  nr=41;rl=0;ru=1;
  r=seq(from=rl,to=ru,by=(ru-rl)/(nr-1));
```

- A temporal interval is defined with nout=21 output points in t, initial and final values of t0=0, tf=200, so that $t = 0, 10, ..., 200$.

```
#
# Temporal grid
  nout=21;t0=0;tf=200;
  times=seq(from=t0,to=tf,by=(tf-t0)/(nout-1));
```

- An IC and history vector u0 is defined for 4*nr=4*21=84 points in r for eqs. (3.3).

```
#
# IC vector
  u0=rep(0,4*nr);
  for(i in 1:nr){
    u0[i]      =fS0(r[i]);
    u0[i+nr]   =fI0(r[i]);
    u0[i+2*nr]=fR0(r[i]);
    u0[i+3*nr]=fV0(r[i]);
  }
  ncall=0;
```

The counter for the calls to pde1a is also initialized.

- The DPDEs for eqs. (3.2) are integrated by the library integrator dede (available in deSolve, [2, Chapter 7]). As expected, the inputs to dede are the IC vector u0, the vector of output values of t, times, and the ODE function, pde1a. The length of u0 (84) informs dede how many ODEs are to be integrated. y, times, func are reserved names.

```
#
# Integration of DPDEs
  uout=dede(y=u0,times=times,func=pde1a);
  nrow(uout);
  ncol(uout);
```

- t is placed in vector tp and $S(r,t), I(r,t), R(r,t), V(r,t)$ from eqs. (3.2) are placed in matrices Sp, Ip, Rp, Vp for numerical and graphical display.

```
#
# Arrays/vectors for ODE solutions
  Sp=matrix(0,nrow=nr,ncol=nout);
  Ip=matrix(0,nrow=nr,ncol=nout);
  Rp=matrix(0,nrow=nr,ncol=nout);
  Vp=matrix(0,nrow=nr,ncol=nout);
  tp=rep(0,nout);
  for(it in 1:nout){
    tp[it]=uout[it,1];
    for( i in 1:nr){
    Sp[i,it]=uout[it,i+1];
    Ip[i,it]=uout[it,i+1+nr];
    Rp[i,it]=uout[it,i+1+2*nr];
    Vp[i,it]=uout[it,i+1+3*nr];
  }
  }
```

- The solutions of eqs. (3.2) are displayed numerically in r and t with two fors.

```
#
# Display numerical solution
  iv=seq(from=1,to=nout,by=5);
  for(it in iv){
    cat(sprintf("\n                    t"));
    cat(sprintf("\n%12.1f\n",tp[it]));
  iv=seq(from=1,to=nr,by=10);
  for(i in iv){
    cat(sprintf("\n                    r"));
    cat(sprintf("\n%12.2f\n",r[i]));
    cat(sprintf("        S(r,t)          I(r,t)"));
    cat(sprintf("\n        R(r,t)          V(r,t)"));
    cat(sprintf("\n%12.2e%12.2e",   Sp[i,it],Ip[i,it]));
    cat(sprintf("\n%12.2e%12.2e\n",Rp[i,it],Vp[i,it]));
  }
  }
```

Every fifth value in *t* and every tenth value in *r* are displayed with
by=5,10.

- The counter for the calls to pde1a is displayed at the end of the solution.

```
#
# Display calls to pde1a
  cat(sprintf("\n ncall = %4d\n",ncall));
```

- The solutions of eqs. (3.2) are plotted in 2D with matplot

```
#
# 2D
  matplot(r,Sp,type="l",xlab="r",ylab="S(r,t)",
          lty=1,main="",lwd=2,col="black");
  matplot(r,Ip,type="l",xlab="r",ylab="I(r,t)",
          lty=1,main="",lwd=2,col="black");
  matplot(r,Rp,type="l",xlab="r",ylab="R(r,t)",
          lty=1,main="",lwd=2,col="black");
  matplot(r,Vp,type="l",xlab="r",ylab="V(r,t)",
          lty=1,main="",lwd=2,col="black");
```

- The solutions of eqs. (3.2) are plotted in 3D with persp.

```
#
# 3D
  persp(r,tp,Sp,theta=60,phi=30,
        xlim=c(rl,ru),ylim=c(t0,tf),xlab="r",
        ylab="t",zlab="S(r,t)");
  persp(r,tp,Ip,theta=60,phi=30,
        xlim=c(rl,ru),ylim=c(t0,tf),xlab="r",
        ylab="t",zlab="I(r,t)");
```

```
persp(r,tp,Rp,theta=60,phi=30,
      xlim=c(rl,ru),ylim=c(t0,tf),xlab="r",
      ylab="t",zlab="R(r,t)");
persp(r,tp,Vp,theta=60,phi=30,
      xlim=c(rl,ru),ylim=c(t0,tf),xlab="r",
      ylab="t",zlab="V(r,t)");
```

This completes the discussion of the main program in Listing 3.1. The DPDE routine, pde1a, is considered next.

3.1.2   DODE routine

DODE routine pde1a follows.

```
  pde1a=function(t,u,parms,lag) {
#
# Function pde1a computes the t derivative
# vectors of S(r,t),I(r,t),R(r,t),V(r,t)
#
# Delayed variable vector
  if (t > tau){
    ulag=lagvalue(t-tau);
  } else {
    ulag=u0;
  }
#
# One vector to four vectors
  S=rep(0,nr);I=rep(0,nr);
  R=rep(0,nr);V=rep(0,nr);
  for(i in 1:nr){
    S[i]=u[i];
    I[i]=u[i+nr];
    R[i]=u[i+2*nr];
    V[i]=u[i+3*nr];
  }
#
# S(r,t-tau), V(r,t-tau)
  Sd=rep(0,nr);Vd=rep(0,nr);
  for(i in 1:nr){
    Sd[i]=ulag[i];
    Vd[i]=ulag[i+3*nr];
  }
#
# Sr,Ir,Rr,Vr
  Sr=dss004(rl,ru,nr,S);
  Ir=dss004(rl,ru,nr,I);
  Rr=dss004(rl,ru,nr,R);
  Vr=dss004(rl,ru,nr,V);
```

```
#
# BCs
  Sr[1]=0;Sr[nr]=0;
  Ir[1]=0;Ir[nr]=0;
  Rr[1]=0;Rr[nr]=0;
  Vr[1]=0;Vr[nr]=0;
#
# Srr,Irr,Rrr,Vrr
  nl=2;nu=2;
  Srr=dss044(rl,ru,nr,S,Sr,nl,nu);
  Irr=dss044(rl,ru,nr,I,Ir,nl,nu);
  Rrr=dss044(rl,ru,nr,R,Rr,nl,nu);
  Vrr=dss044(rl,ru,nr,V,Vr,nl,nu);
#
# DPDEs
  St=rep(0,nr);It=rep(0,nr);
  Rt=rep(0,nr);Vt=rep(0,nr);
  for(i in 1:nr){
    if(i==1){
      St[i]=3*Ds*Srr[i]+Sp-c1*S[i]*V[i]/(1+c2*V[i])-c3*S[i];
      It[i]=3*Di*Irr[i]+c1*Sd[i]*Vd[i]/(1+c2*Vd[i])-c4*I[i];
      Rt[i]=3*Dr*Rrr[i]+c5*I[i]-c6*R[i];
      Vt[i]=3*Dv*Vrr[i]+c7*I[i]+c8*V[i];
    }
    if(i>1){
      ri=2/r[i];
      St[i]=Ds*(Srr[i]+ri*Sr[i])+Sp-c1*S[i]*V[i]/(1+c2*V[i])
            -c3*S[i];
      It[i]=Di*(Irr[i]+ri*Ir[i])+c1*Sd[i]*Vd[i]/(1+c2*Vd[i])
            -c4*I[i];
      Rt[i]=Dr*(Rrr[i]+ri*Rr[i])+c5*I[i]-c6*R[i];
      Vt[i]=Dv*(Vrr[i]+ri*Vr[i])+c7*I[i]+c8*V[i];
    }
  }
#
# Four vectors to one vector
  ut=rep(0,4*nr);
  for(i in 1:nr){
    ut[i]      =St[i];
    ut[i+nr]   =It[i];
    ut[i+2*nr]=Rt[i];
    ut[i+3*nr]=Vt[i];
  }
#
# Increment calls to ncall
  ncall <<- ncall+1;
#
# Return DPDE t vector
  return(list(c(ut)));
```

```
#
# End of pdela
  }
```

Listing 3.2   DPDE routine for eqs. (3.2).

We can note the following details about this listing.

- The function is defined.

```
pdela=function(t,u,parm,lag) {
#
# Function pdela computes the t derivative
# vectors of S(r,t),I(r,t),R(r,t),V(r,t)
```

  t is the current value of *t* in eqs. (3.2). u is the current numerical solution
  to eqs. (3.2). parm is an argument to pass parameters to pdela (unused,
  but required in the argument list). lag is the lag (delay) of the DPDE
  system, which is unused in the current application (the lag is passed to
  pdela from the main program of Listing 3.1 as parameter tau). The arguments must be listed in the order stated to properly interface with dede
  called in the main program of Listing 3.1. The ODE/MOL approxima-
  tions of the derivatives $\dfrac{\partial S(r,t)}{\partial t}, \dfrac{\partial I(r,t)}{\partial t}, \dfrac{\partial R(r,t)}{\partial t}, \dfrac{\partial V(r,t)}{\partial t}$ of eqs. (3.2) are
  calculated and returned to dede as explained subsequently.

- The lagged variables $S(r, t - \tau), V(r, t - \tau)$ in eq. (3.2-2) are computed as
  ulag.

```
#
# Delayed variable vector
  if (t > tau){
    ulag=lagvalue(t-tau);
  } else {
    ulag=u0;
  }
```

  Note the use of the history vector u0, and the lag (delay) tau. u is a 84-
  vector with the four dependent variables of eqs. (3.2) placed according
  to the ICs programmed in the main program of Listing 3.1. For example,
  $S(r, t - \tau)$ is placed in ulag[1] to ulag[21] and $V(r, t - \tau)$ is placed in
  ulag[63] to ulag[84].

- The dependent variable vector, u, is placed in four vectors to facilitate the
  programming of eqs. (3.2).

```
#
# One vector to four vectors
  S=rep(0,nr);I=rep(0,nr);
  R=rep(0,nr);V=rep(0,nr);
```

```
for(i in 1:nr){
  S[i]=u[i];
  I[i]=u[i+nr];
  R[i]=u[i+2*nr];
  V[i]=u[i+3*nr];
}
```

- The delayed variables $S(r, t-\tau)$, $V(r, t-\tau)$ in eq. (3.2-2) are extracted from ulag.

```
#
# S(r,t-tau), V(r,t-tau)
  Sd=rep(0,nr);Vd=rep(0,nr);
  for(i in 1:nr){
    Sd[i]=ulag[i];
    Vd[i]=ulag[i+3*nr];
  }
```

- The first derivatives $\dfrac{\partial S(r,t)}{\partial r}$, $\dfrac{\partial I(r,t)}{\partial r}$, $\dfrac{\partial R(r,t)}{\partial r}$, $\dfrac{\partial V(r,t)}{\partial r}$ are computed with dss004. The arguments of dss004 are explained in Appendix A1.

```
#
# Sr,Ir,Rr,Vr
  Sr=dss004(rl,ru,nr,S);
  Ir=dss004(rl,ru,nr,I);
  Rr=dss004(rl,ru,nr,R);
  Vr=dss004(rl,ru,nr,V);
```

- BCs (3.4) are implemented.

```
#
# BCs
  Sr[1]=0;Sr[nr]=0;
  Ir[1]=0;Ir[nr]=0;
  Rr[1]=0;Rr[nr]=0;
  Vr[1]=0;Vr[nr]=0;
```

Subscripts 1, nr correspond to the boundary values of $\dfrac{\partial S(r,t)}{\partial r}$, $\dfrac{\partial I(r,t)}{\partial r}$, $\dfrac{\partial R(r,t)}{\partial r}$, $\dfrac{\partial V(r,t)}{\partial r}$ at $r = r_l = 0, r = r_u = 1$, respectively.

- The second derivatives $\dfrac{\partial^2 S(r,t)}{\partial r^2}$, $\dfrac{\partial^2 I(r,t)}{\partial r^2}$, $\dfrac{\partial^2 R(r,t)}{\partial r^2}$, $\dfrac{\partial^2 V(r,t)}{\partial r^2}$ in eqs. (3.2) are computed with dss044 (listed and discussed in Appendix A1).

```
#
# Srr,Irr,Rrr,Vrr
```

```
nl=2;nu=2;
Srr=dss044(rl,ru,nr,S,Sr,nl,nu);
Irr=dss044(rl,ru,nr,I,Ir,nl,nu);
Rrr=dss044(rl,ru,nr,R,Rr,nl,nu);
Vrr=dss044(rl,ru,nr,V,Vr,nl,nu);
```

nl=nu=2 specify Neumann BCs (eqs. (3.4)). The arrays Sr,Ir,Br,Vr computed previously provide the boundary values to dss044.

- The MOL programming of eqs. (3.2) steps through the 21 values of *r* in a for.

```
#
# DPDEs
  St=rep(0,nr);It=rep(0,nr);
  Rt=rep(0,nr);Vt=rep(0,nr);
  for(i in 1:nr){
    if(i==1){
      St[i]=3*Ds*Srr[i]+Sp-c1*S[i]*V[i]/(1+c2*V[i])-c3*S[i];
      It[i]=3*Di*Irr[i]+c1*Sd[i]*Vd[i]/(1+c2*Vd[i])-c4*I[i];
      Rt[i]=3*Dr*Rrr[i]+c5*I[i]-c6*R[i];
      Vt[i]=3*Dv*Vrr[i]+c7*I[i]+c8*V[i];
    }
    if(i>1){
      ri=2/r[i];
      St[i]=Ds*(Srr[i]+ri*Sr[i])+Sp-c1*S[i]*V[i]/(1+c2*V[i])
            -c3*S[i];
      It[i]=Di*(Irr[i]+ri*Ir[i])+c1*Sd[i]*Vd[i]/(1+c2*Vd[i])
            -c4*I[i];
      Rt[i]=Dr*(Rrr[i]+ri*Rr[i])+c5*I[i]-c6*R[i];
      Vt[i]=Dv*(Vrr[i]+ri*Vr[i])+c7*I[i]+c8*V[i];
    }
  }
```

Two ifs are used with if(i==1) for *r* = 0 and if(i>1) for *r* > 0 as discussed previously. The correspondence of the DPDEs (eqs. (3.2)) and the programming is a principal feature of the MOL.

- The four derivative vectors St,It,Rt,Vt are placed in one derivative vector, ut, to return to dede (called in the main program of Listing 3.1).

```
#
# Four vectors to one vector
  ut=rep(0,4*nr);
  for(i in 1:nr){
    ut[i]      =St[i];
    ut[i+nr]   =It[i];
    ut[i+2*nr]=Rt[i];
    ut[i+3*nr]=Vt[i];
  }
```

- The counter for the calls to pde1a is incremented and returned to the main program by «-.

```
#
# Increment calls to ncall
  ncall <<- ncall+1;
```

- The derivative vector ut is returned to dede for the next step along the solution.

```
#
# Return DPDE t vector
  return(list(c(ut)));
#
# End of pde1a
  }
```

The derivative ut is returned as a list as required by dede. c is the R vector utility. The final } concludes pde1a.

This completes the discussion of the DPDE routine pde1a. The output from the main program of Listing 3.1 and DPDE routine of Listing 3.2 is considered next.

3.1.3 Numerical, graphical output

Abbreviated output from Listings 3.1 and 3.2 follows in Table 3.2 for ncase=1.

```
[1] 21

[1] 165

            t
          0.0

            r
         0.00
      S(r,t)          I(r,t)
      R(r,t)          V(r,t)
    3.00e+03       1.00e+03
    5.00e+02       1.00e+03

            r
         0.25
      S(r,t)          I(r,t)
      R(r,t)          V(r,t)
    3.00e+03       1.00e+03
    5.00e+02       2.10e+02
```

```
           r
         0.50
     S(r,t)        I(r,t)
     R(r,t)        V(r,t)
   3.00e+03     1.00e+03
   5.00e+02     1.93e+00


           r
         0.75
     S(r,t)        I(r,t)
     R(r,t)        V(r,t)
   3.00e+03     1.00e+03
   5.00e+02     7.81e-04


           r
         1.00
     S(r,t)        I(r,t)
     R(r,t)        V(r,t)
   3.00e+03     1.00e+03
   5.00e+02     1.39e-08
           .            .
           .            .
           .            .
     Output for t=50,100,
          150 removed
           .            .
           .            .
           t
         200.0


           r
         0.00
     S(r,t)        I(r,t)
     R(r,t)        V(r,t)
   4.29e+01     1.99e+02
   8.53e+02     3.02e+03


           r
         0.25
     S(r,t)        I(r,t)
     R(r,t)        V(r,t)
   1.38e+02     2.04e+02
   7.43e+02     7.81e+02


           r
         0.50
     S(r,t)        I(r,t)
     R(r,t)        V(r,t)
   1.22e+03     2.24e+02
   4.66e+02     9.64e+01
```

```
              r
            0.75
       S(r,t)            I(r,t)
       R(r,t)            V(r,t)
      1.29e+03          2.13e+02
      4.50e+02          8.73e+01


              r
            1.00
       S(r,t)            I(r,t)
       R(r,t)            V(r,t)
      1.29e+03          2.13e+02
      4.50e+02          8.73e+01

ncall =   507
```

Table 3.2 Abbreviated output for `ncase=1`

We can note the following details about this output.

- 21 output points in t as the first dimension of the solution matrix `uout` from `dede` as programmed in the main program of Listing 3.1.

- The solution matrix `uout` returned by `dede` has 165 elements as a second dimension. The first element is the value of t. Elements 2–165 in `uout` are $S(r,t), I(r,t), R(r,t), V(r,t)$ for eqs. (3.2) (for each of the 41 output points).

- The solution is displayed for $t = 0, 50, ..., 200$ and $r = 0, 0.25, ..., 1$ as programmed in Listing 3.1.

- ICs (3.3) ($t = 0$) are confirmed. This check is important since if the ICs are not correct, the subsequent solution will also not be correct. Note in particular the Gaussian for $V(r, t = 0)$.

- The solutions are complicated functions of r and t, but generally the four dependent variables $S(r, t), I(r, t), R(r, t), V(r, t)$ appear to be stable (approaching equilibrium or steady state solutions).

- The computational effort is modest, `ncall` = 546, so that `dede` efficiently computed a solution to eqs. (3.2).

The graphical output follows.

Figure 3.1-1 indicates $S(r, t = 0) = 3 \times 10^3$ and the transition in t from this IC.

Figure 3.1-2 indicates $I(r, t = 0) = 1 \times 10^3$ and the transition in t from this IC.

Figure 3.1-3 indicates $R(r, t = 0) = 5 \times 10^2$ and the transition in t from this IC.

Figure 3.1-4 indicates $V(r, t = 0) = 10^3 e^{-25r^2}$ and the transition in t from this IC.

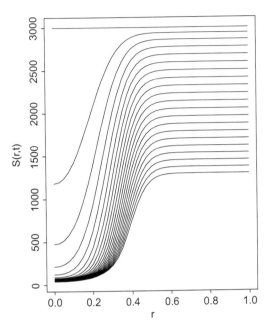

Figure 3.1-1   Numerical solution $S(r,t)$ from eq. (3.2-1), `ncase=1`.

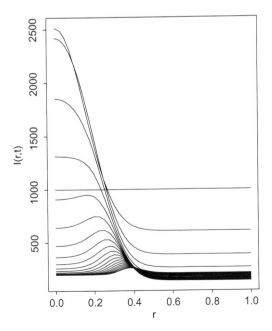

Figure 3.1-2   Numerical solution $I(r,t)$ from eq. (3.2-2), `ncase=1`.

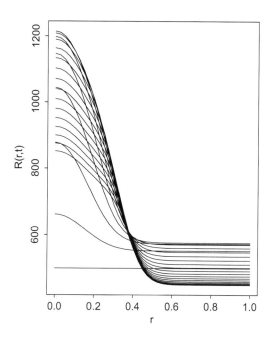

Figure 3.1-3 Numerical solution $R(r, t)$ from eq. (3.2-3), `ncase=1`.

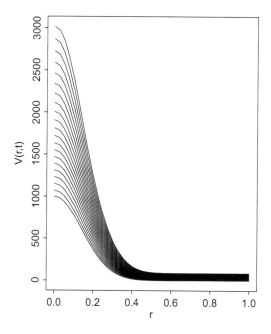

Figure 3.1-4 Numerical solution $V(r, t)$ from eq. (3.2-4), `ncase=1`.

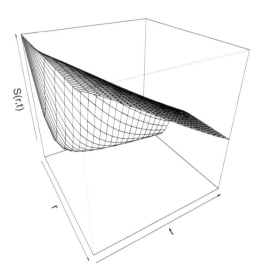

Figure 3.1-5   Numerical solution $S(r, t)$ from eq. (3.2-1), `ncase=1`.

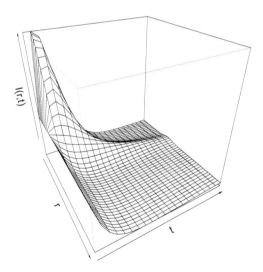

Figure 3.1-6   Numerical solution $I(r, t)$ from eq. (3.2-2), `ncase=1`.

Figure 3.1-5 indicates decreasing $S(r, t)$ with t in response to $V(r, t)$.
Figure 3.1-6 indicates a complex transient in $I(r, t)$ with t.
Figure 3.1-7 indicates a complex transient in $R(r, t)$ with t.
Figure 3.1-8 indicates increasing $V(r, t)$ with t (from $c_8 > 0$ for `ncase=1`).

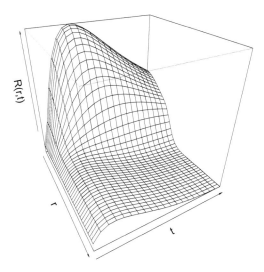

Figure 3.1-7 Numerical solution $R(r,t)$ from eq. (3.2-3), `ncase=1`.

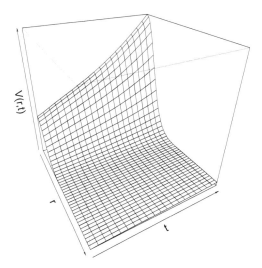

Figure 3.1-8 Numerical solution $V(r,t)$ from eq. (3.2-4), `ncase=1`.

For `ncase=2`, $c_8 < 0$ so that $V(r,t)$ decreases, possibly from vaccination. Abbreviated output for `ncase=2` follows in Table 3.3.

```
[1] 21

[1] 165


        t
    0.0
```

```
           r
         0.00
      S(r,t)        I(r,t)
      R(r,t)        V(r,t)
   3.00e+03     1.00e+03
   5.00e+02     1.00e+03


           r
         0.25
      S(r,t)        I(r,t)
      R(r,t)        V(r,t)
   3.00e+03     1.00e+03
   5.00e+02     2.10e+02


           r
         0.50
      S(r,t)        I(r,t)
      R(r,t)        V(r,t)
   3.00e+03     1.00e+03
   5.00e+02     1.93e+00


           r
         0.75
      S(r,t)        I(r,t)
      R(r,t)        V(r,t)
   3.00e+03     1.00e+03
   5.00e+02     7.81e-04


           r
         1.00
      S(r,t)        I(r,t)
      R(r,t)        V(r,t)
   3.00e+03     1.00e+03
   5.00e+02     1.39e-08
         .            .
         .            .
         .            .
    Output for t=59,100,
         150 removed
         .            .
         .            .
         .            .
           t
         200.0


           r
         0.00
      S(r,t)        I(r,t)
      R(r,t)        V(r,t)
```

```
        2.00e+02      1.69e+02
        8.26e+02      4.32e+02


                 r
            0.25
         S(r,t)        I(r,t)
         R(r,t)        V(r,t)
        5.36e+02      1.46e+02
        6.69e+02      1.26e+02


                 r
            0.50
         S(r,t)        I(r,t)
         R(r,t)        V(r,t)
        1.85e+03      7.35e+01
        3.63e+02      1.86e+01


                 r
            0.75
         S(r,t)        I(r,t)
         R(r,t)        V(r,t)
        1.90e+03      6.87e+01
        3.54e+02      1.70e+01


                 r
            1.00
         S(r,t)        I(r,t)
         R(r,t)        V(r,t)
        1.90e+03      6.87e+01
        3.54e+02      1.70e+01

ncall =   576
```

Table 3.3   Abbreviated output for ncase=2

We can note the following details about this output (since ncase=2 in place of ncase=1 changes only the value of c_8, the preceding discussion of the output for ncase=1 also applies here for ncase=2. The computational effort is modest, ncall = 510, so that dede efficiently computed a solution to eqs. (3.2).

The graphical output follows.

Figure 3.2-1 indicates $S(r, t = 0) = 3 \times 10^3$ and the transition in t from this IC.

Figure 3.2-2 indicates $I(r, t = 0) = 1 \times 10^3$ and the transition in t from this IC.

Figure 3.2-3 indicates $R(r, t = 0) = 5 \times 10^2$ and the transition in t from this IC.

Figure 3.2-4 indicates $V(r, t = 0) = 10^3 e^{-25r^2}$ and the transition in t from this IC.

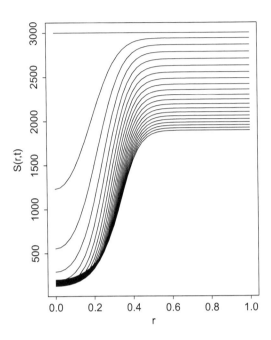

Figure 3.2-1   Numerical solution $S(r, t)$ from eq. (3.2-1), `ncase=2`.

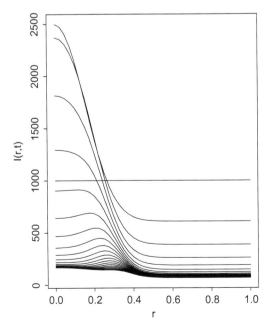

Figure 3.2-2   Numerical solution $I(r, t)$ from eq. (3.2-2), `ncase=2`.

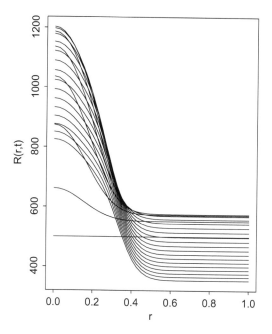

Figure 3.2-3   Numerical solution $R(r, t)$ from eq. (3.2-3), ncase=2.

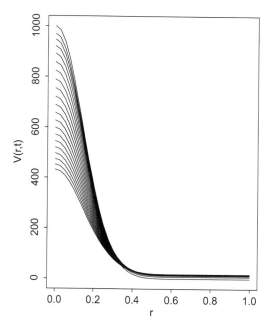

Figure 3.2-4   Numerical solution $V(r, t)$ from eq. (3.2-4), ncase=2.

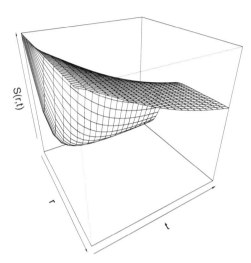

Figure 3.2-5   Numerical solution $S(r, t)$ from eq. (3.2-1), `ncase=2`.

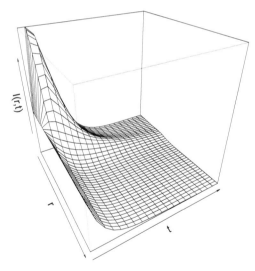

Figure 3.2-6   Numerical solution $I(r, t)$ from eq. (3.2-2), `ncase=2`.

Figure 3.2-5 indicates decreasing $S(r, t)$ with t in response to $V(r, t)$.

Figure 3.2-6 indicates a complex transient in $I(r, t)$ with t.

Figure 3.2-7 indicates a complex transient in $R(r, t)$ with t.

Figure 3.2-8 indicates decreasing $V(r, t)$ with t (from $c_8 < 0$ for `ncase=2`).

In summary, $V(r, t)$ increases for `ncase=1` (with $c_8 > 0$) and decreases for `ncase=2` (with $c_8 < 0$). This could be interpreted, for example, as the effect of vaccination.

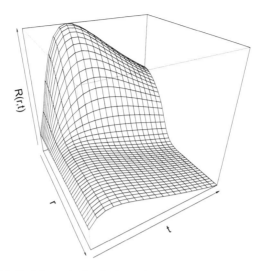

Figure 3.2-7   Numerical solution $R(r, t)$ from eq. (3.2-3), `ncase=2`.

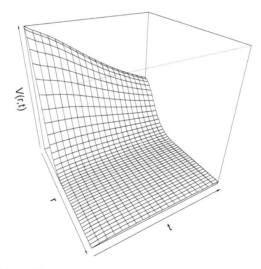

Figure 3.2-8   Numerical solution $V(r, t)$ from eq. (3.2-4), `ncase=2`.

3.2   Summary and conclusions

The 4×4 DPDE model of eqs. (3.1) is implemented in 1D spherical coordinates, eqs. (3.2), with homogeneous Neumann BCs. The effect of the virus IC $V(r, t = 0)$ is investigated within the MOL setting as a system of R routines,

Listings 3.1 and 3.2. This implementation can be used for parametric studies, for example, variation in the delay τ, and variations of the DPDE model.

References

[1] Schiesser, W.E. (2016), *Method of Lines PDE Analysis in Biomedical Science and Engineering*, Wiley, Hoboken, NJ.

[2] Soetaert, K., J. Cash, and F. Mazzia (2012), *Solving Differential Equations in R*, Springer-Verlag, Heidelberg, Germany.

[3] Xu, R., and Z. Ma (2009), An HBV model with diffusion and time delay, *Journal Theoretical Biology*, **257**, pp 499–509.

4

Tumor-Induced Angiogenesis

Introduction

The growth of a tumor is dependent on an adequate input of nutrients, which can be achieved by diffusion when the tumor is small (2–3 mm). When the tumor reaches a critical size, it requires blood circulation for nutrients which is accomplished by the growth of a vascular network (blood vessels, capillaries), usually termed *tumor-induced angiogenesis* (TIA). This process of *tumor neovascularization* from a neighboring blood vessel is described in the following quotation from [2]:

> To ensure its sustained growth, a tumour may secrete chemical compounds which cause neighbouring capillaries to form sprouts which then migrate towards it, furnishing the tumour with an increased supply of nutrients. In this paper a mathematical model is presented which describes the migration of capillary sprouts in response to a chemoattractant field set up by a tumour-released angiogenic factor, sometimes termed a tumour angiogenesis factor (TAF).

In this chapter, delay partial differential equations (DPDEs) model are described for tumor neovascularization spatiotemporal dynamics. The model is coded (programmed) in R, and the chapter concludes with a discussion of the model output.

4.1   DPDE model

The 3×3 (three PDEs in three unknowns) system is listed below with explanations of the PDE terms.

The PDE for the blood vessel density, $u_1(r, t)$, is

$$\frac{\partial u_1(r, t)}{\partial t} = D_1 \left(\frac{\partial^2 u_1(r, t)}{\partial r^2} + \frac{2}{r} \frac{\partial u_1(r, t)}{\partial r} \right) - \gamma u_1(r, t) \qquad (4.1\text{-}1)$$

The terms of eq. (4.1-1) briefly are:

- $\frac{\partial u_1(r,t)}{\partial t}$: Net change of $u_1(r,t)$ with t of the blood vessel density in an incremental volume $4\pi r^2 \Delta r$ where r is the radial position in the tumor external host tissue.

  The incremental volume is the difference in the volumes at $r + \Delta r$ and r:

  $$(4/3)\pi(r+\Delta r)^3 - (4/3)\pi r^3 = (4/3)\pi(r^2 + 2r\Delta r + \Delta r^2)(r + \Delta r) - (4/3)\pi r^3 =$$
  $$(4/3)\pi(r^3 + 2r^2\Delta r + r\Delta r^2 + r^2\Delta r + 2r\Delta r^2 + \Delta r^3) - (4/3)\pi r^3 =$$
  $$(4/3)\pi(3r^2\Delta r + 3r\Delta r^2 + \Delta r^3)$$

  which for small Δr is
  $$4\pi r^2 \Delta r$$

- $D_1\left(\frac{\partial^2 u_1(r,t)}{\partial r^2} + \frac{2}{r}\frac{\partial u_1(r,t)}{\partial r}\right)$: Net rate of diffusion of $u_1(r,t)$ into or out of the incremental volume.

- $-\gamma u_1(r,t)$: Volumetric decay rate of blood vessel density.

The variables and parameters in eq. (4.1-1) are listed in Table 4.1.
  The DPDE for the sprout tip density, $u_2(r,t)$, is

$$\frac{\partial u_2(r,t)}{\partial t} = D_n\left(\frac{\partial^2 u_2(r,t)}{\partial r^2} + \frac{2}{r}\frac{\partial u_2(r,t)}{\partial r}\right) - \chi\frac{1}{r^2}\frac{\partial\left(r^2 u_2(r,t)\frac{\partial u_3(r,t)}{\partial r}\right)}{\partial r}$$
$$+ \alpha_0 u_1(r,t)u_3(r,t-\tau) + \alpha_1 h(u_3(r,t-\tau) - \hat{u}_3)u_2(r,t)u_3(r,t-\tau)$$
$$- \beta u_2(r,t)u_1(r,t) \tag{4.1-2}$$

The terms of eq. (4.1-2) are:

- $\frac{\partial u_2(r,t)}{\partial t}$: Net change of $u_2(r,t)$ with t of the sprout tip density in an incremental volume $4\pi r^2 \Delta r$ where r is the radial position in the tumor external host tissue.

- $D_2\left(\frac{\partial^2 u_2(r,t)}{\partial r^2} + \frac{2}{r}\frac{\partial u_2(r,t)}{\partial r}\right)$: Net rate of diffusion (random motion) of $u_2(r,t)$ into or out of the incremental volume.

- $-\chi\frac{1}{r^2}\frac{\partial\left(r^2 u_2(r,t)\frac{\partial u_3(r,t)}{\partial r}\right)}{\partial r}$: Net rate of chemotaxis motion of $u_2(r,t)$ into or out of the incremental volume. This is a nonlinear term through the multiplication $u_2(r,t)\frac{\partial u_3(r,t)}{\partial r}$ which reflects the effect of the $u_3(r,t)$ gradient on the chemotaxis motion of $u_2(r,t)$.

- $+\alpha_0 u_1(r,t)u_3(r,t-\tau)$: Rate of increase of the tip sprout density, $u_2(r,t)$, from the combined effect of the blood vessel density, $u_1(r,t)$, and delayed TAF $u_3(r,t-\tau)$.

- $+\alpha_1 h(u_3(r,t-\tau)-\hat{u}_3)u_2(r,t)u_3(r,t-\tau)$: Rate of secondary tip proliferation when $u_3(r,t)$ exceeds $\hat{u}_3$, a specified constant (h is the Heaviside step function).

- $-\beta u_2(r,t)u_1(r,t)$: Rate of decay of the sprout tip density, $u_2(r,t)$.

The variables and parameters in eq. (4.1-2) are listed in Table 4.1.
The DPDE for the chemoattractant (TAF) concentration, $u_3(r,t)$, is

$$\frac{\partial u_3(r,t)}{\partial t} = D_c \left(\frac{\partial^2 u_3(r,t)}{\partial r^2} + \frac{2}{r}\frac{\partial u_3(r,t)}{\partial r} \right)$$
$$- \lambda u_3(r,t) - \alpha_1 h(u_3(r,t-\tau)-\hat{u}_3)u_2(r,t)u_3(r,t-\tau) \qquad (4.1\text{-}3)$$

The terms of eq. (4.1-3) are:

- $\dfrac{\partial u_3(r,t)}{\partial t}$: Net change of $u_3(r,t)$ with t of the TAF concentration in an incremental volume $4\pi r^2 \Delta r$ where r is the radial position in the tumor external host tissue.

- $D_3 \left(\dfrac{\partial^2 u_3(r,t)}{\partial r^2} + \dfrac{2}{r}\dfrac{\partial u_3(r,t)}{\partial r} \right)$: Net rate of diffusion (random motion) of $u_3(r,t)$ into or out of the incremental volume.

- $-\lambda u_3(r,t)$: Rate of decay of the TAF concentration, $u_3(r,t)$.

- $-\alpha_1 h(u_3(r,t-\tau)-\hat{u}_3)u_2(r,t)u_3(r,t-\tau)$: Rate of reduction of TAF concentration from secondary tip proliferation when $u_3(r,t)$ exceeds $\hat{u}_3$, a specified constant.

Eqs. (4.1) are variations of eqs. (8), (9), (10), [2] and are designated the TIA model.

The variables and parameters of eqs. (4.1) are explained in Table 4.1.
The added features of eqs. (4.1) are:

- In accordance with the usual convention of naming dependent variables in PDE models,[1] the following names are used (in place of the names in [2]):

$$\rho(r,t) = u_1(r,t); \quad n(r,t) = u_2(r,t); \quad c(r,t) = u_3(r,t)$$

---

[1] The generic name u_n denotes a PDE dependent variable with number (index) n. For the present model, $n = 1, 2, 3$ correspond to $u_1(r,t), u_2(r,t), u_3(r,t)$.

Variable, parameter	Description
$u_1(r,t)$	blood vessel density
$u_2(r,t)$	sprout tip density
$u_3(r,t)$	chemoattractant (TAF) concentration
r	radial position
t	time
D_1	coefficient of $u_1(r,t)$ random motion
D_2	coefficient of $u_2(r,t)$ random motion
D_3	coefficient of $u_3(r,t)$ random motion
χ	coefficient of $u_2(r,t)$ chemotaxis motion
γ	decay rate coefficient for $u_1(r,t)$
λ	$u_3(r,t)$ natural decay rate constant
α_0	rate constant for appearance of tips
α_1	rate constant for secondary tip proliferation
h	Heaviside function that switches at $\hat{u}_3$
$\hat{u}_3$	$u_3(r,t)$ that triggers secondary tip proliferation
β	rate constant for tip annihilation
τ	delay in sprout tip formation from TAF

Table 4.1   Variables and parameters of TIA model

- The tumor spatial geometry is represented with the independent variable r that is the 1D radial component of spherical coordinates (r, θ, ϕ) (the angular components are neglected and a single spatial r is used[2]). A 1D spherical coordinate is considered as a better representation of the tumor and surrounding volume (external host tissue) than the 1D Cartesian coordinate used in [2].

- Diffusion has been included in all three PDEs, with diffusivities D_1, D_2, D_3. In particular, the convection of [2, eq. (8)], has been replaced by diffusion with D_1.

- A delay τ is used to represent the time lag of the sprout formation from the TAF in the tip density DPDE, eq. (4.1-2).

---

[2]A single spatial region is used here to simplify the TIA model and maintain a basic, introductory discussion of the model. Two spatial regions for the tumor and the external host tissue (extending to the limbus) could be included by either (1) using two concentric spheres, or (2) varying the properties spatially within this proposed single region model.

If Cartesian coordinates are preferred, the incremental volume would be $\Delta x \Delta y \Delta z$ and the radial groups would not have the second term, for example, $D_1 \left(\dfrac{\partial^2 u_1(r,t)}{\partial r^2} + \dfrac{2}{r} \dfrac{\partial u_1(r,t)}{\partial r} \right)$ would become $D_1 \left(\dfrac{\partial^2 u_1(x,t)}{\partial x^2} \right)$.

Eqs. (4.1) are second order in r so they each require two BCs.

$$\frac{\partial u_1(r = r_l, t)}{\partial r} = 0; \quad u_1(r = r_u, t) = u_{1L}(1 - e^{-kt}) \tag{4.2-1,2}$$

$$\frac{\partial u_2(r = r_l, t)}{\partial r} = 0; \quad u_2(r = r_u, t) = u_{2L}(1 - e^{-kt}) \tag{4.2-3,4}$$

$$u_3(r = r_l, t) = u_{30}(1 - e^{-kt}); \quad \frac{\partial u_3(r = r_u, t)}{\partial r} = 0 \tag{4.2-5,6}$$

where r_u is the r region outer boundary and $u_{1L}, u_{2L}, u_{30}, k$ are parameters (constants) to be specified.

Eqs. (4.1) are first order in t so each requires one initial condition (IC).

$$u_1(r, t = 0) = 0 \tag{4.3-1}$$
$$u_2(r, t = 0) = 0 \tag{4.3-2}$$
$$u_3(r, t = 0) = 0 \tag{4.3-3}$$

The R routines that implement eqs. (4.1), (4.2), (4.3) follow, starting with the main program.

4.1.1 Main program

The main program is in the following Listing 4.1.

```
#
# Three PDE model
#
# Delete previous workspaces
  rm(list=ls(all=TRUE))
#
# Access ODE integrator
  library("deSolve");
#
# Access functions for numerical solution
  setwd("f:/dpde/chap4");
  source("pde1a.R");
  source("dss004.R");
  source("dss044.R");
  source("h.R");
#
# Parameters
  D1=1.0e-02;
  D2=1.0e-02;
  D3=1.0e-02;
```

```
  chi=0.4;
  gamma=0.25;
  lambda=1;
  alpha0=50;
  alpha1=10;
  u3sec=0.2;
  beta=5;
  tau=1;
  u1L=1;
  u2L=1;
  u30=1;
  k=1;
#
# Grid in x
  rl=0.2;ru=1;nr=41;dr=(ru-rl)/(nr-1);
  r=seq(from=rl,to=ru,by=dr);
#
# Initial condition
  u0=rep(0,(3*nr));
  for(i in 1:nr){
    u0[i]     =0;
    u0[i+nr]  =0;
    u0[i+2*nr]=0;
  }
#
# Interval in t
  t0=0;tf=2;nout=21;
  tout=seq(from=t0,to=tf,by=(tf-t0)/(nout-1));
  ncall=0;
#
# Integration of DPDEs
  out=dede(y=u0,times=tout,func=pde1a);
  nrow(out);
  ncol(out);
#
# Store solution
  u1=matrix(0,nrow=nr,ncol=nout);
  u2=matrix(0,nrow=nr,ncol=nout);
  u3=matrix(0,nrow=nr,ncol=nout);
  t=rep(0,nout);
  for(it in 1:nout){
  for(i in 1:nr){
    u1[i,it]=out[it,i+1];
    u2[i,it]=out[it,i+1+nr];
    u3[i,it]=out[it,i+1+2*nr];
    t[it]=out[it,1];
  }
  }
#
```

```
# Display numerical solution
  iv=seq(from=1,to=nout,by=5);
  for(it in iv){
  cat(sprintf(
    "\n\n          t          r     u1(r,t)"));
  cat(sprintf(
    "\n          t          r     u2(r,t)"));
  cat(sprintf(
    "\n          t          r     u3(r,t)"));
  iv=seq(from=1,to=nr,by=5);
  for(i in iv){
    cat(sprintf("\n%9.2e%11.2e%12.4f",
      t[it],r[i],u1[i,it]));
    cat(sprintf("\n%9.2e%11.2e%12.4f",
      t[it],r[i],u2[i,it]));
    cat(sprintf("\n%9.2e%11.2e%12.4f\n",
      t[it],r[i],u3[i,it]));
  }
  }
#
# Display ncall
  cat(sprintf("\n\n ncall = %2d",ncall));
#
# Plot numerical solutions
#
# 2D
  matplot(r,u1,type="l",xlab="r",ylab="u1(r,t)",
          lty=1,main="",lwd=2,col="black");
  matplot(r,u2,type="l",xlab="r",ylab="u2(r,t)",
          lty=1,main="",lwd=2,col="black");
  matplot(r,u3,type="l",xlab="r",ylab="u3(r,t)",
          lty=1,main="",lwd=2,col="black");
#
# 3D
  persp(r,t,u1,theta=60,phi=45,
        xlim=c(rl,ru),ylim=c(t0,tf),xlab="r",ylab="t",
        zlab="u1(r,t)");
  persp(r,t,u2,theta=60,phi=45,
        xlim=c(rl,ru),ylim=c(t0,tf),xlab="r",ylab="t",
        zlab="u2(r,t)");
  persp(r,t,u3,theta=60,phi=45,
        xlim=c(rl,ru),ylim=c(t0,tf),xlab="r",ylab="t",
        zlab="u3(r,t)");
```

Listing 4.1   Main program for eqs. (4.1), (4.2), (4.3).

The following discussion of Listing 4.1 partly repeats the discussion of Listing 3.1 in Chapter 3, but is given so this discussion of the TIA model is self-contained.

- Previous workspaces are deleted.

```
#
# Three DPDE model
#
# Delete previous workspaces
  rm(list=ls(all=TRUE))
```

- The R ODE integrator library deSolve is accessed. Then the directory
  with the files for the solution of eqs. (4.1) is designated. Note that setwd
  (set working directory) uses / rather than the usual \.

```
#
# Access ODE integrator
  library("deSolve");
#
# Access functions for numerical solution
  setwd("f:/dpde/chap4");
  source("pde1a.R");
  source("dss004.R");
  source("dss044.R");
  source("h.R");
```

pde1a.R is the routine for eqs. (4.1) (discussed subsequently) based on
the method of lines (MOL), a general algorithm for PDEs [5]. dss004,
dss044 are library routines for the calculation of first and second spatial derivatives. These routines are listed in Appendix A1 with additional
explanation.

- The parameters of eqs. (4.1) are specified. These values are taken in part
  from Byrne and Chaplain [2, Figure 1]. Values not available from this reference are specified to give reasonable numerical solutions to eqs. (4.1),
  (4.2), (4.3). A discussion of the parameter value selection is given in
  Appendix A4.

```
#
# Parameters
  D1=1.0e-02;
  D2=1.0e-02;
  D3=1.0e-02;
  chi=0.4;
  gamma=0.25;
  lambda=1;
  alpha0=50;
  alpha1=10;
  u3sec=0.2;
  beta=5;
  tau=1;
  u1L=1;
  u2L=1;
```

```
u30=1;
k=1;
```

The parameter names are defined/explained in Table 4.1.

- A grid in r is defined for $r_l = 0.2 \le r \le r_u = 1$ with 41 points so $r = 0.20, 0.22, ..., 1$.

```
#
# Grid in r
  rl=0.2;ru=1;nr=41;dr=(ru-rl)/(nr-1);
  r=seq(from=rl,to=ru,by=dr);
```

- An IC and history vector u0 is defined for 3*nr=3*41=123 points in r for eqs. (4.3).

```
#
# Initial condition
  u0=rep(0,(3*nr));
  for(i in 1:nr){
    u0[i]      =0;
    u0[i+nr]   =0;
    u0[i+2*nr]=0;
  }
```

- A temporal interval is defined with nout=21 output points in t, initial and final values of t0=0, tf=2, so that $t = 0, 0.1, ..., 2$.

```
#
# Interval in t
  t0=0;tf=2;nout=21;
  tout=seq(from=t0,to=tf,by=(tf-t0)/(nout-1));
  ncall=0;
```

- The MOL/ODEs for eqs. (4.1) are integrated by the library integrator dede (available in deSolve, [6, Chapter 7]). As expected, the inputs to dede are the IC vector u0, the vector of output values of t, times, and the ODE function, pde1a. The length of u0 (123) informs dede how many DODEs are to be integrated. y, times, func are reserved names. The counter for the calls to pde1a is also initialized.

```
#
# Integration of DPDEs
  out=dede(y=u0,times=tout,func=pde1a);
  nrow(out);
  ncol(out);
```

- t is placed in vector t and $u_1(r, t), u_2(r, t), u_3(r, t)$ from eqs. (4.1) are placed in matrices u1, u2, u3 for numerical and graphical display.

```
#
# Store solution
  u1=matrix(0,nrow=nr,ncol=nout);
  u2=matrix(0,nrow=nr,ncol=nout);
  u3=matrix(0,nrow=nr,ncol=nout);
  t=rep(0,nout);
  for(it in 1:nout){
  for(i in 1:nr){
    u1[i,it]=out[it,i+1];
    u2[i,it]=out[it,i+1+nr];
    u3[i,it]=out[it,i+1+2*nr];
    t[it]=out[it,1];
  }
  }
```

- The solutions of eqs. (4.1) are displayed numerically in r and t with two fors.

```
#
# Display numerical solution
  iv=seq(from=1,to=nout,by=5);
  for(it in iv){
  cat(sprintf(
    "\n\n            t              r       u1(r,t)"));
  cat(sprintf(
    "\n          t              r       u2(r,t)"));
  cat(sprintf(
    "\n          t              r       u3(r,t)"));
  iv=seq(from=1,to=nr,by=5);
  for(i in iv){
    cat(sprintf("\n%9.2e%11.2e%12.4f",
      t[it],r[i],u1[i,it]));
    cat(sprintf("\n%9.2e%11.2e%12.4f",
      t[it],r[i],u2[i,it]));
    cat(sprintf("\n%9.2e%11.2e%12.4f\n",
      t[it],r[i],u3[i,it]));
  }
  }
```

Every fifth value in r and t is displayed with by=5.

- The counter for the calls to pde1a is displayed at the end of the solution.

```
#
# Display ncall
  cat(sprintf("\n\n ncall = %2d",ncall));
```

- The solutions $u_1(r,t), u_2(r,t), u_3(r,t)$ are plotted in 2D with matplot.

```
#
# Plot numerical solutions
#
# 2D
  matplot(r,u1,type="l",xlab="r",ylab="u1(r,t)",
          lty=1,main="",lwd=2,col="black");
  matplot(r,u2,type="l",xlab="r",ylab="u2(r,t)",
          lty=1,main="",lwd=2,col="black");
  matplot(r,u3,type="l",xlab="r",ylab="u3(r,t)",
          lty=1,main="",lwd=2,col="black");
```

- The solutions $u_1(r,t), u_2(r,t), u_3(r,t)$ are plotted in 3D with persp.

```
#
# 3D
  persp(r,t,u1,theta=60,phi=45,
        xlim=c(rl,ru),ylim=c(t0,tf),xlab="r",ylab="t",
        zlab="u1(r,t)");
  persp(r,t,u2,theta=60,phi=45,
        xlim=c(rl,ru),ylim=c(t0,tf),xlab="r",ylab="t",
        zlab="u2(r,t)");
  persp(r,t,u3,theta=60,phi=45,
        xlim=c(rl,ru),ylim=c(t0,tf),xlab="r",ylab="t",
        zlab="u3(r,t)");
```

This completes the discussion of the main program in Listing 4.1. The ODE/MOL routine pde1a is considered next.

4.1.2 DODE routine

```
  pde1a=function(t,u,parms){
#
# Function pde1a computes the t derivative
# vector of u1(r,t), u2(r,t), u3(r,t)
#
# Delayed variable vector
  if (t > tau){
    ulag=lagvalue(t-tau);
  } else {
    ulag=u0;
  }
#
# One vector to three vectors
  u1=rep(0,nr);
  u2=rep(0,nr);
  u3=rep(0,nr);
  for(i in 1:nr){
    u1[i]=u[i];
```

```
    u2[i]=u[i+nr];
    u3[i]=u[i+2*nr];
  }
#
# Dirichlet BCs
  u1[nr]=u1L*(1-exp(-k*t));
  u2[nr]=u2L*(1-exp(-k*t));
  u3[1] =u30*(1-exp(-k*t));
#
# u1r,nr,cr
  u1r=dss004(rl,ru,nr,u1);
  u2r=dss004(rl,ru,nr,u2);
  u3r=dss004(rl,ru,nr,u3);
#
# Neumann BCs
  u1r[1] =0;
  u2r[1] =0;
  u3r[nr]=0;
#
# u1rr,nrr,crr
  nl=2;nu=1;
  u1rr=dss044(rl,ru,nr,u1,u1r,nl,nu);
  nl=2;nu=1;
  u2rr=dss044(rl,ru,nr,u2,u2r,nl,nu);
  nl=1;nu=2;
  u3rr=dss044(rl,ru,nr,u3,u3r,nl,nu);
#
# u3(r,t-tau)
  u3d=rep(0,nr);
  for(i in 1:nr){
    u3d[i]=ulag[i+2*nr];
  }
#
# DPDEs, u1t(r,t),u2t(r,t),u3t(r,t)
  u1t=rep(0,nr);
  u2t=rep(0,nr);
  u3t=rep(0,nr);
  for(i in 1:nr){
    ri=2/r[i];
    u1t[i]=D1*(u1rr[i]+ri*u1r[i])-gamma*u1[i];
    term1=alpha1*h(u3d[i]-u3sec)*u2[i]*u3d[i];
    u2t[i]=D2*(u2rr[i]+ri*u2r[i])-chi*(u2[i]*u3rr[i]+
           u2r[i]*u3r[i]+ri*u2[i]*u3r[i])+
           alpha0*u1[i]*u3d[i]+term1-beta*u2[i]*u1[i];
    u3t[i]=D3*(u3rr[i]+ri*u3r[i])-lambda*u3[i]-term1;
  }
#
# Three vectors to one vector
  ut=rep(0,(3*nr));
```

```
  for(i in 1:nr){
    ut[i]      =u1t[i];
    ut[i+nr]   =u2t[i];
    ut[i+2*nr]=u3t[i];
  }
#
# Increment calls to pde1a
  ncall<<-ncall+1;
#
# Return derivative vector
  return(list(c(ut)));
}
```

Listing 4.2   DODE routine for eqs. (4.1), (4.2), (4.3).

We can note the following details about this listing.
- The function is defined.

```
  pde1a=function(t,u,parm){
#
# Function pde1a computes the t derivative
# vector of u1(r,t), u2(r,t), u3(r,t)
```

t is the current value of *t* in eqs. (4.1). u is the current numerical solution to eqs. (4.1). parm is an argument to pass parameters to pde1a (unused, but required in the argument list). lag is the lag (delay) of the DPDE system, which is unused in the current application (the lag is passed to pde1a from the main program of Listing 4.1 as parameter tau). The arguments must be listed in the order stated to properly interface with dede called in the main program of Listing 4.1. The DPDE approximations of the derivatives $\dfrac{\partial u_1(r,t)}{\partial t}$, $\dfrac{\partial u_2(r,t)}{\partial t}$, $\dfrac{\partial u_3(r,t)}{\partial t}$ of eqs. (4.1) are calculated and returned to dede as explained subsequently.

- The lagged variable $u_3(r, t - \tau)$ in eqs. (4.2-2,3) computed as ulag.

```
#
# Delayed variable vector
  if (t > tau){
    ulag=lagvalue(t-tau);
  } else {
    ulag=u0;
  }
```

u (second input argument of pde1a) is a 123-vector with the three dependent variables of eqs. (4.1) placed according to the ICs programmed in the main program of Listing 4.1. ulag has the lagged values of u, so that $u_3(r, t-\tau)$ is placed in ulag[83] to ulag[123]. u0, the IC vector, is also the history vector when t <= tau.

- The dependent variable vector, u, is placed in three vectors to facilitate the programming of eqs. (4.1).

```
#
# One vector to three vectors
  u1=rep(0,nr);
  u2=rep(0,nr);
  u3=rep(0,nr);
  for(i in 1:nr){
    u1[i]=u[i];
    u2[i]=u[i+nr];
    u3[i]=u[i+2*nr];
  }
```

- Dirichlet BCs (4.2-2), (4.2-4), (4.2-5) are programmed.

```
#
# Dirichlet BCs
  u1[nr]=u1L*(1-exp(-k*t));
  u2[nr]=u2L*(1-exp(-k*t));
  u3[1] =u30*(1-exp(-k*t));
```

Subscripts 1, nr correspond to $r = r_l, r_u$, respectively.

- The delayed variable $u_3(r, t - \tau)$ in eqs. (4.1-2,3) is extracted from ulag.

```
#
# u3(r,t-tau)
  u3d=rep(0,nr);
  for(i in 1:nr){
    u3d[i]=ulag[i+2*nr];
  }
```

- The first derivatives $\dfrac{\partial u_1(r,t)}{\partial r}$, $\dfrac{\partial u_2(r,t)}{\partial r}$, $\dfrac{\partial u_3(r,t)}{\partial r}$ are computed with dss004. The arguments of dss004 are explained in Appendix A1.

```
#
# u1r,nr,cr
  u1r=dss004(rl,ru,nr,u1);
  u2r=dss004(rl,ru,nr,u2);
  u3r=dss004(rl,ru,nr,u3);
```

- Neumann BCs (4.2-1), (4.2-3), (4.2-6) are applied.

```
#
# Neumann BCs
  u1r[1] =0;
  u2r[1] =0;
  u3r[nr]=0;
```

Subscripts 1, nr correspond to $r = r_l, r_u$, respectively.

- The second derivatives $\dfrac{\partial^2 u_1(r,t)}{\partial r^2}, \dfrac{\partial^2 u_2(r,t)}{\partial r^2}, \dfrac{\partial^2 u_3(r,t)}{\partial r^2}$ in eq. (4.1) are computed with `dss044` (listed and discussed in Appendix A1).

```
#
# u1rr,nrr,crr
  nl=2;nu=1;
  u1rr=dss044(rl,ru,nr,u1,u1r,nl,nu);
  nl=2;nu=1;
  u2rr=dss044(rl,ru,nr,u2,u2r,nl,nu);
  nl=1;nu=2;
  u3rr=dss044(rl,ru,nr,u3,u3r,nl,nu);
```

For eqs. (4.1-1), (4.1-2), the BC at $r = r_l$ is Neumann (nl=2) and at $r = r_u$ it is Dirichlet (nu=1). For eq. (4.1-3), the BC at $r = r_l$ is Dirichlet (nl=1) and at $r = r_u$ it is Neumann (nu=2).

- The delayed variable $u_3(r, t - \tau)$ in eqs. (4.2-2,3) is extracted from `ulag`.

```
#
# u3(r,t-tau)
  u3d=rep(0,nr);
  for(i in 1:nr){
    u3d[i]=ulag[i+2*nr];
  }
```

- The MOL programming of eq. (4.1) steps through the 41 values of r in a `for`.

```
#
# DPDEs, u1t(r,t),u2t(r,t),u3t(r,t)
  u1t=rep(0,nr);
  u2t=rep(0,nr);
  u3t=rep(0,nr);
  for(i in 1:nr){
    ri=2/r[i];
    u1t[i]=D1*(u1rr[i]+ri*u1r[i])-gamma*u1[i];
    term1=alpha1*h(u3d[i]-u3sec)*u2[i]*u3d[i];
    u2t[i]=D2*(u2rr[i]+ri*u2r[i])-chi*(u2[i]*u3rr[i]+
             u2r[i]*u3r[i]+ri*u2[i]*u3r[i])+
             alpha0*u1[i]*u3d[i]+term1-beta*u2[i]*u1[i];
    u3t[i]=D3*(u3rr[i]+ri*u3r[i])-lambda*u3[i]-term1;
  }
```

Branching for $r = 0$ is not required since the grid in r is $0.2 \le r \le 1$ ($r = 0$ is not included). The correspondence of the DPDEs (eqs. (4.1)) and the programming indicates an important feature of the MOL.

- The three derivative vectors `u1t,u2t,u3t` are placed in one derivative vector, `ut`, to return to `dede` (called in the main program of Listing 4.1).

```
#
# Three vectors to one vector
  ut=rep(0,(3*nr));
  for(i in 1:nr){
    ut[i]      =u1t[i];
    ut[i+nr]   =u2t[i];
    ut[i+2*nr]=u3t[i];
  }
```

- The counter for the calls to pde1a is incremented and returned to the main program by «-.

```
#
# Increment calls to pde1a
  ncall<<-ncall+1;
```

- The derivative vector ut is returned to dede for the next step along the solution.

```
#
# Return derivative vector
  return(list(c(ut)));
}
```

The derivative ut is returned as a list as required by dede. c is the R vector utility. The final } concludes pde1a.

4.1.3 Subordinate routine

The Heaviside function h in eqs. (4.1-2,3) is listed next.

```
  h=function(t){
#
# Function h implements the Heaviside
# unit step function
#
# h(t)
  if(t<0 ){h=0};
  if(t>=0){h=1};
#
# Return h
  return(c(h));
}
```

Listing 4.3   Heaviside function h in eqs. (4.1-2,3).

This function is straightforward and does not require additional explanation.

This completes the discussion of the DODE routine pde1a. The output from the main program of Listing 4.1 and DODE routine of Listing 4.2 is considered next.

4.1.4   Numerical, graphical output

Abbreviated output from Listings 4.1–4.3 follows in Table 4.2.

```
[1]  21

[1]  124
```

t	r	u1(r,t)
t	r	u2(r,t)
t	r	u3(r,t)
0.00e+00	2.00e-01	0.0000
0.00e+00	2.00e-01	0.0000
0.00e+00	2.00e-01	0.0000
0.00e+00	3.00e-01	0.0000
0.00e+00	3.00e-01	0.0000
0.00e+00	3.00e-01	0.0000
0.00e+00	4.00e-01	0.0000
0.00e+00	4.00e-01	0.0000
0.00e+00	4.00e-01	0.0000
0.00e+00	5.00e-01	0.0000
0.00e+00	5.00e-01	0.0000
0.00e+00	5.00e-01	0.0000
0.00e+00	6.00e-01	0.0000
0.00e+00	6.00e-01	0.0000
0.00e+00	6.00e-01	0.0000
0.00e+00	7.00e-01	0.0000
0.00e+00	7.00e-01	0.0000
0.00e+00	7.00e-01	0.0000
0.00e+00	8.00e-01	0.0000
0.00e+00	8.00e-01	0.0000
0.00e+00	8.00e-01	0.0000
0.00e+00	9.00e-01	0.0000
0.00e+00	9.00e-01	0.0000
0.00e+00	9.00e-01	0.0000
0.00e+00	1.00e+00	0.0000
0.00e+00	1.00e+00	0.0000
0.00e+00	1.00e+00	0.0000
.	.	
.	.	
.	.	

```
        Output for t=0.5,1.0,1.5
                 removed

            .                    .
            .                    .
            .                    .
            t            r        u1(r,t)
            t            r        u2(r,t)
            t            r        u3(r,t)
  2.00e+00    2.00e-01      0.0001
  2.00e+00    2.00e-01      0.0003
  2.00e+00    2.00e-01      0.7996

  2.00e+00    3.00e-01      0.0002
  2.00e+00    3.00e-01      0.0006
  2.00e+00    3.00e-01      0.1845

  2.00e+00    4.00e-01      0.0011
  2.00e+00    4.00e-01      0.0015
  2.00e+00    4.00e-01      0.0405

  2.00e+00    5.00e-01      0.0051
  2.00e+00    5.00e-01      0.0049
  2.00e+00    5.00e-01      0.0084

  2.00e+00    6.00e-01      0.0199
  2.00e+00    6.00e-01      0.0149
  2.00e+00    6.00e-01      0.0016

  2.00e+00    7.00e-01      0.0655
  2.00e+00    7.00e-01      0.0375
  2.00e+00    7.00e-01      0.0003

  2.00e+00    8.00e-01      0.1812
  2.00e+00    8.00e-01      0.0811
  2.00e+00    8.00e-01      0.0000

  2.00e+00    9.00e-01      0.4264
  2.00e+00    9.00e-01      0.2005
  2.00e+00    9.00e-01      0.0000

  2.00e+00    1.00e+00      0.8646
  2.00e+00    1.00e+00      0.7255
  2.00e+00    1.00e+00      0.0000

ncall = 3037
```

Table 4.2   Abbreviated output from Listings 4.1–4.3

We can note the following details about this output.

- 21 output points in t as the first dimension of the solution matrix `uout` from `dede` as programmed in the main program of Listing 4.1.

- The solution matrix `uout` returned by `dede` has 124 elements as a second dimension. The first element is the value of t. Elements 2–124 in `uout` are $u_1(r,t)$, $u_2(r,t)$, $u_3(r,t)$ for eqs. (4.1) (for each of the 21 output points).

- The solution is displayed for $t = 0, 0.5, ..., 2$ and $r = 0.2, 0.4, ..., 1$ as programmed in Listing 4.1.

- Homogeneous ICs (4.1-3) are confirmed (at $t = 0$).

- The computational effort is manageable, `ncall = 3027`, so that `dede` efficiently computed a solution to eqs. (4.1).

The details of the solutions are presented in Figures 4.1-1,2,3 as 2D and in Figures 4.1-4,5,6 as 3D.

These figures indicate the effect of the Dirichlet and Neumann BCs (4.2) (the movement of the solutions away from the homogeneous ICs (4.3)). Of particular interest is the development of the neovascularization $(u_1(r,t), u_2(r,t))$ in response to the TAF $(u_3(r,t))$.

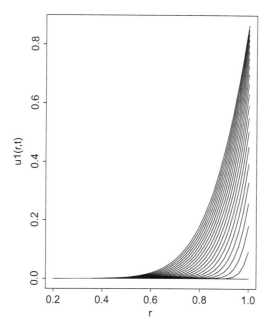

Figure 4.1-1 Numerical solution $u_1(r,t)$ from eq. (4.1-1), `matplot`.

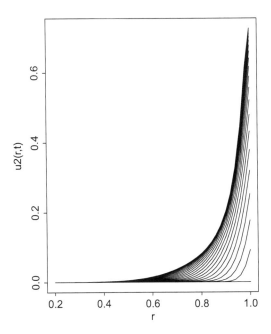

Figure 4.1-2   Numerical solution $u_2(r, t)$ from eq. (4.1-2), `matplot`.

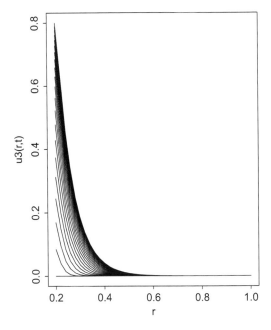

Figure 4.1-3   Numerical solution $u_3(r, t)$ from eq. (4.1-3), `matplot`.

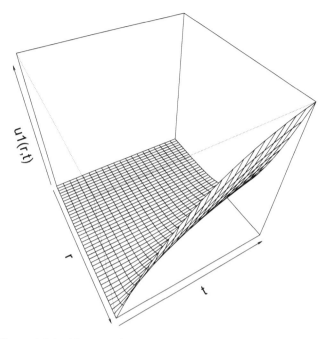

Figure 4.1-4 Numerical solution $u_1(r,t)$ from eq. (4.1-1), persp.

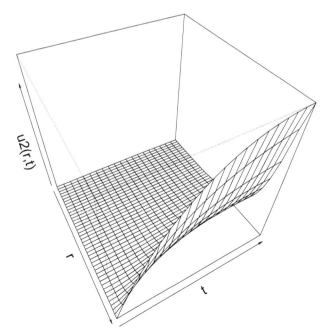

Figure 4.1-5 Numerical solution $u_2(r,t)$ from eq. (4.1-2), persp.

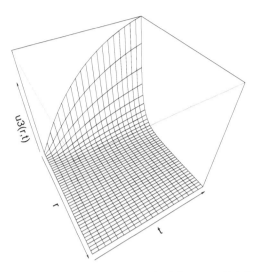

Figure 4.1-6   Numerical solution $u_3(r,t)$ from eq. (4.1-3), `persp`.

As a further investigation with the model, the penetration of the profiles can be increased, for example, by increasing the diffusivities from `1.0e-02` to `1.0e-01` (in the main program of Listing 4.1). This is left as an exercise.

4.2   Summary and conclusions

The MOL implementation of the 3×3 DPDE model of eqs. (4.1), (4.2), (4.3) is straightforward as implemented in the R routines of Listings 4.1–4.3. Of particular interest is the inclusion of a lagged variable, $u_3(r, t - \tau)$, to account for the delayed effect of the TAF in TIA (neovascularization).

Appendix A4: Implementation of the model

This appendix details a suggested approach to coding a PDE model such as eqs. (4.1), (4.2), (4.3). If the entire model, with all of its details, is programmed, it almost certainly will not compile and execute. The problem with this approach is that the reason for the failure to execute probably will not be clear (even with error messages from the compiler).

Rather, experience has indicated that an incremental approach is the best way to proceed by adding small additional details/features. At each step the coding can be tested (compiled and executed). If at a particular step the execution fails, the source of the error(s) most likely will be in the last step (last details added), and it can then be identified and corrected.

This incremental approach is illustrated by the development of the R routines in Listings 4.1–4.3. To start, the RHSs of eqs. (4.1) are set to zero in `pde1a` of Listing 4.2.

```
#
# DPDEs, u1t(r,t),u2t(r,t),u3t(r,t)
  u1t=rep(0,nr);
  u2t=rep(0,nr);
  u3t=rep(0,nr);
  for(i in 1:nr){
    u1t[i]=0;
    u2t[i]=0;
    u3t[i]=0;
  }
```

Listing A4.1   ODE/MOL derivatives set to zero.

The test of Listing A4.1 is worth carrying out for four reasons:

- The syntax is checked by the R compiler. The R error messages are complete and comprehensive so that correction of syntax errors is usually straightforward. In particular, missing elements such as parameter values will be identified.

- The dimensions of the solution matrix can be checked. In the present case, these dimensions are

  ```
  [1]  21
  ```

  ```
  [1]  124
  ```

  These dimensions indicate 21 points in t and $3(41) + 1 = 124$ solution values.

- The ICs can be checked which is important since they are the starting point of the solution. If the ICs are incorrect, the subsequent solution will be incorrect.

- The solution should remain at the ICs. If the solution departs from the ICs, a programming error is indicated, probably in the DODE routine. The IC check can be completed before going on to the next step of programming development.

The RHS derivatives can then be programmed for a case for which a known (and stable) solution is expected. For example, if just diffusion is programmed, we would expect the solution to vary smoothly in r and t.

```
#
# DPDEs, u1t(r,t),u2t(r,t),u3t(r,t)
  u1t=rep(0,nr);
  u2t=rep(0,nr);
  u3t=rep(0,nr);
  for(i in 1:nr){
    ri=2/r[i];
    u1t[i]=D1*(u1rr[i]+ri*u1r[i]);
    u2t[i]=D2*(u2rr[i]+ri*u2r[i]);
    u3t[i]=D3*(u3rr[i]+ri*u3r[i]);
  }
```

Listing A4.2   DPDEs with only diffusion.

The `matplot` 2D graphical output from Listing A4.2 follows in Figures A4.1.

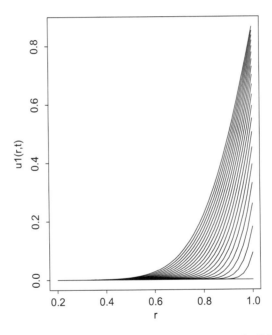

Figure A4.1-1   Numerical solution $u_1(r,t)$ from eq. (4.1-1), diffusion only.

The solutions have a smooth transition from the homogeneous ICs. Also, the interval in r can be checked ($0.2 \leq r \leq 1$) and the BCs can be checked at least qualitatively. For example, the homogeneous Neumann (zero slope) BCs for $u_1(r = r_l = 0.2, t)$, $u_2(r = r_l = 0.2, t)$, $u_3(r = r_u = 1, t)$ are clear. Further, the t-dependent Dirichlet BCs indicate increasing values with t as expected from $1 - e^{-kt}$ (the ICs are consistent with the BCs (i.e., both are zero) so a discontinuous (step) change does not occur from the ICs to the BCs).

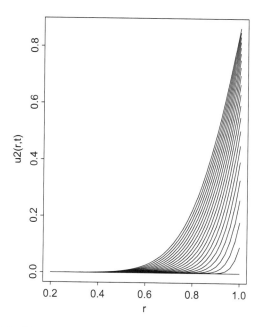

Figure A4.1-2 Numerical solution $u_2(r,t)$ from eq. (4.1-2), diffusion only.

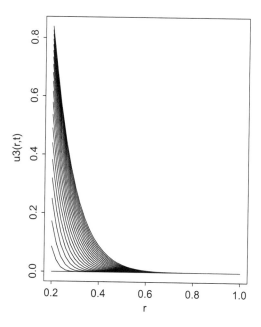

Figure A4.1-3 Numerical solution $u_3(r,t)$ from eq. (4.1-3), diffusion only.

These observations were used to select the numerical values of the diffusivities (1.0e-02) to give a significant/smooth variation of the solutions in r.

The level of computation can also be checked. For Listing A4.2, this is ncall=1206 which is modest so that dede in Listing 4.1 computed a solution efficiently.

The next step will be to add additional RHS terms from eqs. (4.1). This should be done incrementally, and if the compilation/execution fails at a particular step, the additions for this step are most likely the cause of the failure.

In the present case, some of the densities/concentrations eventually became negative (physically impossible). Some experimentation with the parameters (the parameters were changed incrementally from zero to the values indicated in Listing 4.1) identified beta as a cause of the negative solutions. When it was changed from 50 (taken from Byrne and Chaplain [2]) to 5, the problem of negative solutions was resolved. In general, the RHS derivatives cannot remain negative for too long or the solutions will drop below zero (even though the Dirichlet BC values are increasing in t), as for example, from $-\beta u_2(r,t)u_1(r,t)$ in eq. (4.1-2).

The preceding discussion indicates that the implementation of a new model is typically a trial-and-error procedure. Incremental additions generally are the required approach since including all of the details at the beginning will usually lead to failed executions with no apparent cause.

To conclude, DODE/DPDEs are frequently applied in biology/chemistry/physics which requires consideration of physical units. The dependent variables typically represent cell densities or biochemical concentrations. The units for these components must be selected to be consistent throughout the system of equations. Alternatively, the dependent variables can be normalized with respect to selected base values and therefore have representative values near one.

The independent variables are space and time, which are specified on the basis of the analyst's understanding of the problem system. In the case of eqs. (4.1), the outer tumor boundary is at $r = r_l = 0.2$ cm. The time scale is selected as $t_0 = 0 \leq t \leq t_f = 2$ months (mo). The diffusivities therefore have the units 0.01 cm/mo^2 or $0.01/((60)(60)(24)(30)) = 3.86 \times 10^{-9}$ cm^2/s, which is a representative value for cells and large biochemical molecules.

However, the selection of units for the dependent and independent variables can be investigated through experimentation/execution of the computer routines. The final selection warrants explanation as part of the documentation for the model routines.

References

[1] Anderson, A.R.A., and M.A.J. Chaplain (1998), Continuous and discrete mathematical models of tumor-induced angiogenesis, *Bulletin of Mathematical Biology* **60**, pp 857–900.

[2] Byrne, H.M., and M.A.J. Chaplain (1995), Mathematical models for tumour angiogenesis: Numerical simulations and nonlinear wave solution, *Bulletin of Mathematical Biology* **57**, pp 461–486.

[3] Gupta, M.K., and R.-Y. Qin (2003), Mechanism and its regulation of tumor-induced angiogenesis, *World Journal of Gastroenterology*, **9**, pp 1144–1155.

[4] Mantzaris, N.V., S. Webb, and H. G. Othmer (2004), Mathematical modeling of tumor-induced angiogenesis, *Journal of Mathematical Biology*, **49**, pp 111–187.

[5] Schiesser, W.E. (2016), *Method of Lines PDE Analysis in Biomedical Science and Engineering*, Wiley, Hoboken, NJ.

[6] Soetaert, K., J. Cash, and F. Mazzia (2012), *Solving Differential Equations in R*, Springer-Verlag, Heidelberg, Germany.

5

Metastatic Cancer Cell Invasion of Tissue

Introduction

In this chapter, the metastatic invasion of tissue by cancer cells is analyzed with a 5×5 (five equations in five unknowns) system of delay partial differential equations (DPDEs). The origin and spatiotemporal evolution of the DPDE dependent variables is described in the following statement from Andasari et al. [1]:

> The ability of cancer cells to break out of tissue compartments and invade locally gives solid tumours a defining deadly characteristic. One of the first steps of invasion is the remodelling of the surrounding tissue or extra cellular matrix (ECM) and a major part of this process is the over-expression of proteolytic enzymes, such as the urokinase-type plasminogen activator (uPA) and matrix metalloproteinases (MMPs), by the cancer cells to break down ECM proteins. Degradation of the matrix enables the cancer cells to migrate through the tissue and subsequently to spread to secondary sites in the body, a process known as metastasis. In this paper we undertake an analysis of a mathematical model of cancer cell invasion of tissue, or ECM, which focuses on the role of the urokinase plasminogen activation system. The model consists of a system of five reaction–diffusion–taxis partial differential equations describing the interactions between cancer cells, uPA, uPA inhibitors, plasmin, and the host tissue. Cancer cells react chemotactically and haptotactically to the spatio-temporal effects of the uPA system.

5.1 DPDE model

The following 5×5 DPDE model is a variant of the model reported in [1, 2].

$$\frac{\partial n}{\partial t} = D_n \nabla^2 n - \nabla \cdot \left[\chi_n n \nabla u + \chi_p n \nabla p + \chi_v n \nabla v \right] + \mu_1 n (1 - n) \qquad (5.1\text{-}1)$$

$$\frac{\partial v}{\partial t} = D_v \nabla^2 v - \delta v m + \phi_{21} u p - \phi_{22} v p + \mu_2 v (1 - v) \qquad (5.1\text{-}2)$$

$$\frac{\partial u}{\partial t} = D_u \nabla^2 u - \phi_{31} pu - \phi_{33} n(t-\tau)u + \alpha_{31} n(t-\tau) \qquad (5.1\text{-}3)$$

$$\frac{\partial p}{\partial t} = D_p \nabla^2 p - \phi_{41} pu - \phi_{42} pv + \alpha_{41} m \qquad (5.1\text{-}4)$$

$$\frac{\partial m}{\partial t} = D_m \nabla^2 m + \phi_{52} pu + \phi_{53} n(t-\tau)u - \phi_{54} m \qquad (5.1\text{-}5)$$

where the dependent variables are

Variable	Interpretation
n	cancer cell density
$n(t-\tau)$	delayed cancer cell density
v	VN (ECM protein vitronectin) concentration
u	uPA (urokinase plasminogen activator) concentration
p	PAI-1 (plasminogen activator inhibitor type-1) concentration
m	plasmin (matrix degrading enzyme) concentration

Table 5.1   Dependent variables of PDEs (5.1)

A schematic diagram of the model of eqs. (5.1) is given in [1, Fig. 1], that indicates the relationships between the dependent variables of Table 5.1. Eqs. (5.1) are in a coordinate-free format as listed in Table 5.2.

Variable	Interpretation
$\nabla \cdot$	vector divergence operator
∇	scalar gradient operator
$\dfrac{\partial}{\partial t}$	rate (partial derivative in time, t)

Table 5.2   Model differential operators

If eqs. (5.1) are expressed in 1D spherical coordinates, that is, r in (r, θ, ϕ), they are

$$\frac{\partial n}{\partial t} = D_n \frac{1}{r^2} \frac{\partial \left(r^2 n \frac{\partial n}{\partial r} \right)}{\partial r} - \chi_u \frac{1}{r^2} \frac{\partial \left(r^2 n \frac{\partial u}{\partial r} \right)}{\partial r}$$

$$- \chi_p \frac{1}{r^2} \frac{\partial \left(r^2 n \frac{\partial p}{\partial r} \right)}{\partial r} - \chi_v \frac{1}{r^2} \frac{\partial \left(r^2 n \frac{\partial v}{\partial r} \right)}{\partial r}$$

$$+ \mu_1 n (1 - n)$$

$$= D_n \left(\frac{\partial^2 n}{\partial r^2} + \frac{2}{r} \frac{\partial n}{\partial r} \right) - \chi_u \left(n \frac{\partial^2 u}{\partial r^2} + \frac{\partial n}{\partial r} \frac{\partial u}{\partial r} + \frac{2}{r} n \frac{\partial u}{\partial r} \right)$$

$$- \chi_p \left(n \frac{\partial^2 p}{\partial r^2} + \frac{\partial n}{\partial r} \frac{\partial p}{\partial r} + \frac{2}{r} n \frac{\partial p}{\partial r} \right) - \chi_v \left(n \frac{\partial^2 v}{\partial r^2} + \frac{\partial n}{\partial r} \frac{\partial v}{\partial r} + \frac{2}{r} n \frac{\partial v}{\partial r} \right)$$

$$+ \mu_1 n (1 - n) \tag{5.2-1}$$

$$\frac{\partial v}{\partial t} = D_v \frac{1}{r^2} \frac{\partial \left(r^2 \frac{\partial v}{\partial r} \right)}{\partial r} - \delta v m + \phi_{21} u p - \phi_{22} v p + \mu_2 v (1 - v)$$

$$= D_v \left(\frac{\partial^2 v}{\partial r^2} + \frac{2}{r} \frac{\partial v}{\partial r} \right) - \delta v m + \phi_{21} u p - \phi_{22} v p + \mu_2 v (1 - v) \tag{5.2-2}$$

$$\frac{\partial u}{\partial t} = D_u \frac{1}{r^2} \frac{\partial \left(r^2 \frac{\partial u}{\partial r} \right)}{\partial r} - \phi_{31} p u - \phi_{33} n(t - \tau) u + \alpha_{31} n(t - \tau)$$

$$= D_u \left(\frac{\partial^2 u}{\partial r^2} + \frac{2}{r} \frac{\partial u}{\partial r} \right) - \phi_{31} p u - \phi_{33} n(t - \tau) u + \alpha_{31} n(t - \tau) \tag{5.2-3}$$

$$\frac{\partial p}{\partial t} = D_p \frac{1}{r^2} \frac{\partial \left(r^2 \frac{\partial p}{\partial r} \right)}{\partial r} - \phi_{41} p u - \phi_{42} p v + \alpha_{41} m$$

$$= D_p \left(\frac{\partial^2 p}{\partial r^2} + \frac{2}{r} \frac{\partial p}{\partial r} \right) - \phi_{41} p u - \phi_{42} p v + \alpha_{41} m \tag{5.2-4}$$

$$\frac{\partial m}{\partial t} = D_m \frac{1}{r^2} \frac{\partial \left(r^2 \frac{\partial m}{\partial r} \right)}{\partial r} + \phi_{52} p u + \phi_{53} n(t - \tau) u - \phi_{54} m$$

$$= D_m \left(\frac{\partial^2 m}{\partial r^2} + \frac{2}{r} \frac{\partial m}{\partial r} \right) + \phi_{52} p u + \phi_{53} n(t - \tau) u - \phi_{54} m \tag{5.2-5}$$

Eqs. (5.2) are the starting point for the subsequent computer implementation and analysis.

To complete the model specification, eqs. (5.2) are second order in r and each requires two boundary conditions (BCs).

$$\frac{\partial n(r = r_l, t)}{\partial r} = k_n(n(r = r_l, t) - n_{cc}) \tag{5.3-1}$$

$$\frac{\partial v(r = r_l, t)}{\partial r} = k_v(v(r = r_l, t) - v_{cc}) \tag{5.3-2}$$

$$\frac{\partial u(r = r_l, t)}{\partial r} = k_u(u(r = r_l, t) - u_{cc}) \tag{5.3-3}$$

$$\frac{\partial p(r = r_l, t)}{\partial r} = k_p(p(r = r_l, t) - p_{cc}) \tag{5.3-4}$$

$$\frac{\partial m(r = r_l, t)}{\partial r} = k_m(m(r = r_l, t) - m_{cc}) \tag{5.3-5}$$

Robin BCs (5.3) specify continuity in the flux of $n(r, t), v(r, t), u(r, t), p(r, t),$ $m(r, t)$ at the cell outer boundary, $r = r_l$, respectively. k_n to k_m are mass transfer coefficients. n_{cc} is the cell density at the tumor outer boundary. $v_{cc}, u_{cc}, p_{cc}, m_{cc}$ are the concentrations of $VN, uPA, PAI - 1$ and plasmin (matrix degrading enzyme) at the tumor outer boundary, respectively.

$$\frac{\partial n(r = r_u, t)}{\partial r} = 0 \tag{5.4-1}$$

$$\frac{\partial v(r = r_u, t)}{\partial r} = 0 \tag{5.4-2}$$

$$\frac{\partial u(r = r_u, t)}{\partial r} = 0 \tag{5.4-3}$$

$$\frac{\partial p(r = r_u, t)}{\partial r} = 0 \tag{5.4-4}$$

$$\frac{\partial m(r = r_u, t)}{\partial r} = 0 \tag{5.4-5}$$

Neumann BCs (5.4) specify no flux at the tissue (ECM) outer boundary, $r = r_u$.

Eqs. (5.2) are first order in t and each requires one initial condition (IC).

$$n(r, t = 0) = n_0(r) \tag{5.5-1}$$
$$v(r, t = 0) = v_0(r) \tag{5.5-2}$$
$$u(r, t = 0) = u_0(r) \tag{5.5-3}$$
$$p(r, t = 0) = p_0(r) \tag{5.5-4}$$
$$m(r, t = 0) = m_0(r) \tag{5.5-5}$$

where $n_0(r)$ to $m_0(r)$ are the functions to be specified.

Parameter	Value
D_n, D_v, D_u	0.35, 0, 0.25
D_p, D_m	0.35, 0.491
χ_u, χ_p, χ_v	0.305, 0.375, 0.285
δ, μ_1, μ_2	8.15, 0.25, 0.15
ϕ_{21}, ϕ_{22}	0.75, 0.55
ϕ_{31}, ϕ_{33}	0.75, 0.3
ϕ_{41}, ϕ_{42}	0.75, 0.55
ϕ_{52}, ϕ_{53}, ϕ_{54}	0.0011, 0.0075, 0.005
α_{31}, α_{41}	0.215, 0.5
τ	10
n_{cc}, v_{cc}, u_{cc}	1, 1, 1
p_{cc}, m_{cc}	1, 1
k_n, k_v, k_u	1, 1, 1
k_p, k_m	1, 1

Table 5.3 Parameters of PDEs (5.2) (from Andasari et al. [1, p. 147], with revisions)

The numerical values of the parameters (constants) in eqs. (5.2) and (5.3) are variants of the values in [1] and additions for BCs (5.3).

Some details concerning the parameter values in Table 5.3 follow, with reference to [1].

- D_n, D_u, D_p, D_m: Increased by 10^2 or 10^3 to increase the spatial variation in r (eqs. (5.2-1,3,4,5)).

- D_v: Added to eq. (5.2-2) to provide radial diffusion in $v(r,t)$ (eq. (5.2-2)).

- χ_u, χ_p, χ_v: Increased by 10 to increase the chemotaxis, haptotaxis (eq. (5.2-1)).

- ϕ_{52}, ϕ_{53}, ϕ_{54}: Decreased by 10^{-2} to decrease $m(r,t)$ (eq. (5.2-5)).

- n_{cc}, v_{cc}, u_{cc} p_{cc}, m_{cc}: Added for BCs (5.3).

- k_n, k_v, k_u, k_p, k_m: Added for BCs (5.3).

To complete this discussion of the model, the terms in eqs. (5.2) are described briefly.

Eq. (5.2-1) is a balance on the cancer cell density, $n(r,t)$, for an incremental volume $4\pi r^2 \Delta r$, followed by $\Delta r \to 0$, where r is the radial position in the external tissue.

- $\dfrac{\partial n(r,t)}{\partial t}$: Increase (term > 0) or decrease (term < 0) of $n(r,t)$ with t.

- $D_n \left(\dfrac{\partial^2 n}{\partial r^2} + \dfrac{2}{r} \dfrac{\partial n}{\partial r} \right)$: Net rate of diffusion (random motion) of $n(r,t)$ into or out of the incremental volume.

- $-\chi_u \dfrac{1}{r^2} \dfrac{\partial \left(r^2 n \dfrac{\partial u}{\partial r} \right)}{\partial r}$: Rate of chemotaxis from the uPA $= u(r,t)$ gradient and cancer cell density nonlinear coefficient into or out of the incremental volume.

- $-\chi_p \dfrac{1}{r^2} \dfrac{\partial \left(r^2 n \dfrac{\partial p}{\partial r} \right)}{\partial r}$: Rate of chemotaxis from the PAI-1 $= p(r,t)$ gradient and cancer cell density nonlinear coefficient into or out of the incremental volume.

- $-\chi_v \dfrac{1}{r^2} \dfrac{\partial \left(r^2 n \dfrac{\partial v}{\partial r} \right)}{\partial r}$: Rate of haptotaxis from the VN $= v(r,t)$ gradient and cancer cell density nonlinear coefficient into or out of the incremental volume.

- $+\mu_1 n(1 - n)$: Logistic rate of cancer cell proliferation (production) in the incremental volume.

All of these terms are nondimensional rates with space normalized by $L = 0.1$ cm, time normalized by $L^2/D = 10^4$ s, and cancer cell density normalized by $n_{d0} = 6.7 \times 10^7$ cell/cm^3 [1].

Application of the MOL to the RHS of eq. (5.2-1) gives a system of ODEs in t from the LHS which can be solved numerically with a library ODE integrator to give the numerical $n(r,t)$.

Eq. (5.2-2) is a balance on the ECM protein vitronectin, with concentration $v(r,t)$, for an incremental volume $4\pi r^2 \Delta r$, followed by $\Delta r \to 0$, where r is the radial position in the external tissue.

- $\dfrac{\partial v(r,t)}{\partial t}$: Increase (term > 0) or decrease (term < 0) of $v(r,t)$ with t.

- $D_v \left(\dfrac{\partial^2 v}{\partial r^2} + \dfrac{2}{r} \dfrac{\partial v}{\partial r} \right)$: Net rate of diffusion (random motion) of VM (ECM protein vitronectin) with concentration $v(r,t)$ into or out of the incremental volume.

- $-\delta vm$: Degradation of VM by plasmin.

- $+\phi_{21} up$: uPA and PAI-1 production interaction.

- $-\phi_{22}vp$: VM and PAI-1 depletion interaction.

- $+\mu_2 v(1 - v)$: VM remodeling (reformation).

Again, eq. (5.2-2) is integrated within the MOL format.

Eq. (5.2-3) is a balance on the uPA (urokinase plasminogen activator), with concentration $u(r,t)$, for an incremental volume $4\pi r^2 \Delta r$, followed by $\Delta r \to 0$, where r is the radial position in the external tissue.

- $\dfrac{\partial u}{\partial t}$: Increase (term > 0) or decrease (term < 0) of $u(r,t)$ with t.

- $D_u \left(\dfrac{\partial^2 u}{\partial r^2} + \dfrac{2}{r} \dfrac{\partial u}{\partial r} \right)$: Net rate of diffusion (random motion) of uPA (urokinase plasminogen activator) with concentration $u(r,t)$ into or out of the incremental volume.

- $-\phi_{31}pu$: PAI-1 and uPA depletion interaction.

- $-\phi_{33}n(t - \tau)u$: Delayed cancer cell density and uPA interaction. The delayed variable is used to reflect the lag in the depletion of uPA by cancer cells.

- $+\alpha_{31}n(t - \tau)$: Delayed uPA production. The delayed variable is used to reflect the lag in the production of uPA stimulated by cancer cells.

Eq. (5.2-3) is integrated within the MOL format.

Eq. (5.2-4) is a balance on the PAI-1 (plasminogen activator inhibitor type-1), with concentration $p(r,t)$, for an incremental volume $4\pi r^2 \Delta r$, followed by $\Delta r \to 0$, where r is the radial position in the external tissue.

- $\dfrac{\partial p}{\partial t}$: Increase (term > 0) or decrease (term < 0) of $p(r,t)$ with t.

- $D_p \left(\dfrac{\partial^2 p}{\partial r^2} + \dfrac{2}{r} \dfrac{\partial p}{\partial r} \right)$: Net rate of diffusion (random motion) of PAI-1 (plasminogen activator inhibitor type-1) with concentration $p(r,t)$ into or out of the incremental volume.

- $-\phi_{41}pu$: PAI-1 and uPA depletion interaction.

- $-\phi_{42}pv$: PAI-1 and VM depletion interaction.

- $+\alpha_{41}m$: PAI-1 production by matrix degrading enzyme.

Eq. (5.2-4) is integrated within the MOL format.

Eq. (5.2-5) is a balance on the plasmin (matrix degrading enzyme), with concentration $m(r,t)$, for an incremental volume $4\pi r^2 \Delta r$, followed by $\Delta r \to 0$, where r is the radial position in the external tissue.

- $\dfrac{\partial m}{\partial t}$: Increase (term > 0) or decrease (term < 0) of $m(r,t)$ with t.

- $D_m \left(\dfrac{\partial^2 m}{\partial r^2} + \dfrac{2}{r} \dfrac{\partial m}{\partial r} \right)$: Net rate of diffusion (random motion) of plasmin (matrix degrading enzyme) with concentration $m(r,t)$ into or out of the incremental volume.

- $+\phi_{52} pu$: PAI-1 and uPA production interaction.

- $+\phi_{53} n(t - \tau)u$: Delayed cancer cell and uPA production interaction.

- $-\phi_{54} m$: ECM degradation.

Eq. (5.2-5) is integrated within the MOL format.

Eqs. (5.2), (5.3), (5.4), and (5.5), with the information in Tables 5.1 and 5.3 constitute the DPDE model. The computer implementation of the model is considered next.

5.1.1 Main program

The main program for eqs. (5.2), (5.3), (5.4), and (5.5) is in Listing 5.1.

```
#
# Five DPDE model
#
# Delete previous workspaces
  rm(list=ls(all=TRUE))
#
# Access ODE integrator
  library("deSolve");
#
# Access functions for numerical solution
  setwd("f:/dpde/chap5");
  source("pde1a.R");
  source("dss004.R");
  source("dss044.R");
#
# Parameters
  Dn=3.5e-01; Dv=0; Du=2.5e-01;
  Dp=3.5e-01; Dm=4.91e-01;
  chiu=3.05e-01; chip=3.75e-01; chiv=2.85e-01;
  delta=8.15; mu1=0.25; mu2=0.15;
  phi21=0.75; phi22=0.55;
  phi31=0.75;  phi33=0.3;
```

```
  phi41=0.75; phi42=0.55;
  phi52=0.11e-02; phi53=0.75e-02; phi54=0.5e-02;
  alpha31=0.215; alpha41=0.5;
  tau=10;
  ncc=1; vcc=1; ucc=1; pcc=1; mcc=1;
  kn=1;   kv=1;   ku=1;   kp=1;   km=1;
#
# Grid in r
  rl=1;ru=10;ng=51;
  r=seq(from=rl,to=ru,by=(ru-rl)/(ng-1));
#
# Initial condition
  u0=rep(0,(5*ng));
  for(i in 1:ng){
    u0[i]       =0;
    u0[i+ng]   =0;
    u0[i+2*ng]=0;
    u0[i+3*ng]=0;
    u0[i+4*ng]=0;
  }
#
# Interval in t
  t0=0;tf=100;nout=26;
  tout=seq(from=t0,to=tf,by=(tf-t0)/(nout-1));
  ncall=0;
#
# ODE integration
  out=dede(y=u0,times=tout,func=pde1a,
           atol=1.0e-04);
  nrow(out)
  ncol(out)
#
# Store solution
  n=matrix(0,nrow=ng,ncol=nout);
  v=matrix(0,nrow=ng,ncol=nout);
  u=matrix(0,nrow=ng,ncol=nout);
  p=matrix(0,nrow=ng,ncol=nout);
  m=matrix(0,nrow=ng,ncol=nout);
  t=rep(0,nout);
  for(it in 1:nout){
  for(i in 1:ng){
    n[i,it]=out[it,i+1];
    v[i,it]=out[it,i+1+ng];
    u[i,it]=out[it,i+1+2*ng];
    p[i,it]=out[it,i+1+3*ng];
    m[i,it]=out[it,i+1+4*ng];
    t[it]=out[it,1];
  }
  }
```

```
#
# Display numerical solution
  iv=seq(from=1,to=nout,by=5);
  for(it in iv){
  cat(sprintf(
    "\n\n          t              r        n(r,t)"));
  cat(sprintf(
    "\n         t            r        v(r,t)"));
  cat(sprintf(
    "\n        t            r        u(r,t)"));
  cat(sprintf(
    "\n         t            r        p(r,t)"));
  cat(sprintf(
    "\n         t            r        m(r,t)"));
  iv=seq(from=1,to=ng,by=5);
  for(i in iv){
    cat(sprintf("\n%9.2e%11.2e%12.4f",
      t[it],r[i],n[i,it]));
    cat(sprintf("\n%9.2e%11.2e%12.4f",
      t[it],r[i],v[i,it]));
    cat(sprintf("\n%9.2e%11.2e%12.4f",
      t[it],r[i],u[i,it]));
    cat(sprintf("\n%9.2e%11.2e%12.4f",
      t[it],r[i],p[i,it]));
    cat(sprintf("\n%9.2e%11.2e%12.4f\n",
      t[it],r[i],m[i,it]));
  }
  }
#
# Display ncall
  cat(sprintf("\n\n ncall = %2d",ncall));
#
# Plot numerical solutions
#
# 2D
  matplot(r,n,type="l",xlab="r",ylab="n(r,t)",
          lty=1,main="",lwd=2,col="black");
  matplot(r,v,type="l",xlab="r",ylab="v(r,t)",
          lty=1,main="",lwd=2,col="black");
  matplot(r,u,type="l",xlab="r",ylab="u(r,t)",
          lty=1,main="",lwd=2,col="black");
  matplot(r,p,type="l",xlab="r",ylab="p(r,t)",
          lty=1,main="",lwd=2,col="black");
  matplot(r,m,type="l",xlab="r",ylab="m(r,t)",
          lty=1,main="",lwd=2,col="black");
#
# 3D
  persp(r,t,n,theta=45,phi=45,
        xlim=c(rl,ru),ylim=c(t0,tf),xlab="r",ylab="t",
```

```
        zlab="n(r,t)");
persp(r,t,v,theta=45,phi=45,
        xlim=c(rl,ru),ylim=c(t0,tf),xlab="r",ylab="t",
        zlab="v(r,t)");
persp(r,t,u,theta=45,phi=45,
        xlim=c(rl,ru),ylim=c(t0,tf),xlab="r",ylab="t",
        zlab="u(r,t)");
persp(r,t,p,theta=60,phi=45,
        xlim=c(rl,ru),ylim=c(t0,tf),xlab="r",ylab="t",
        zlab="p(r,t)");
persp(r,t,m,theta=60,phi=45,
        xlim=c(rl,ru),ylim=c(t0,tf),xlab="r",ylab="t",
        zlab="m(r,t)");
```

Listing 5.1   Main program for eqs. (5.2), (5.3), (5.4) and (5.5).

We can note the following details about Listing 5.1.

- Previous workspaces are deleted.

```
#
# Five DPDE model
#
# Delete previous workspaces
  rm(list=ls(all=TRUE))
```

- The R ODE integrator library deSolve is accessed. Then the directory with the files for the solution of eqs. (5.2) is designated. Note that setwd (set working directory) uses / rather than the usual \.

```
#
# Access ODE integrator
  library("deSolve");
#
# Access functions for numerical solution
  setwd("f:/dpde/chap5");
  source("pde1a.R");
  source("dss004.R");
  source("dss044.R");
```

pde1a.R is the routine for eqs. (5.2) (discussed subsequently) based on the method of lines (MOL), a general algorithm for PDEs [3]. dss004, dss044 are library routines for the calculation of first and second spatial derivatives. These routines are listed in Appendix A1 with additional explanation.

- The parameters of eqs. (5.2) as listed in Table 5.3 are programmed.

```
#
# Parameters
  Dn=3.5e-01; Dv=0; Du=2.5e-01;
```

```
Dp=3.5e-01; Dm=4.91e-01;
chiu=3.05e-01; chip=3.75e-01; chiv=2.85e-01;
delta=8.15; mu1=0.25; mu2=0.15;
phi21=0.75; phi22=0.55;
phi31=0.75;  phi33=0.3;
phi41=0.75; phi42=0.55;
phi52=0.11e-02; phi53=0.75e-02; phi54=0.5e-02;
alpha31=0.215; alpha41=0.5;
tau=10;
ncc=1; vcc=1; ucc=1; pcc=1; mcc=1;
kn=1;   kv=1;   ku=1;   kp=1;   km=1;
```

- A grid in r is defined for $r_l = 1 \le r \le r_u = 10$ with 51 points so $r = 1, (10-1)/50, ..., 10$. The value of ng was selected as large enough to avoid gridding effects in r, but small enough to avoid excessively large values of ncall.

```
#
# Grid in r
  rl=1;ru=10;ng=51;
  r=seq(from=rl,to=ru,by=(ru-rl)/(ng-1));
```

- An IC and history vector u0 is defined for 5*ng=5*51=255 points in r for eqs. (5.5).

```
#
# Initial condition
  u0=rep(0,(5*ng));
  for(i in 1:ng){
     u0[i]      =0;
     u0[i+ng]   =0;
     u0[i+2*ng]=0;
     u0[i+3*ng]=0;
     u0[i+4*ng]=0;
  }
```

- A temporal interval is defined with nout=26 output points in t, initial and final values of t0=0, tf=100, so that $t = 0, 4, ..., 100$.

```
#
# Interval in t
  t0=0;tf=100;nout=26;
  tout=seq(from=t0,to=tf,by=(tf-t0)/(nout-1));
  ncall=0;
```

The counter for the calls to pde1a is also initialized.

- The MOL/ODEs for eqs. (5.2) are integrated by the library integrator dede (available in deSolve, [5, Chapter 7]). As expected, the inputs to dede are the IC vector u0, the vector of output values of t, times, and the ODE function, pde1a. The length of u0 (255) informs dede how many DODEs are to be integrated. y, times, func are reserved names.

```
#
# ODE integration
  out=dede(y=u0,times=tout,func=pde1a,
           atol=1.0e-04);
  nrow(out);
  ncol(out);
```

The error tolerance atol was increased from the default 1.0e-06 to 1.0e-04 to reduce the value of ncall (determined by trial and error).

- t is placed in vector t and $n(r,t)$, $v(r,t)$, $u(r,t)$, $p(r,t)$, $m(r,t)$ from eqs. (5.2) are placed in matrices n, v, u, p, m for numerical and graphical display.

```
#
# Store solution
  n=matrix(0,nrow=ng,ncol=nout);
  v=matrix(0,nrow=ng,ncol=nout);
  u=matrix(0,nrow=ng,ncol=nout);
  p=matrix(0,nrow=ng,ncol=nout);
  m=matrix(0,nrow=ng,ncol=nout);
  t=rep(0,nout);
  for(it in 1:nout){
  for(i in 1:ng){
    n[i,it]=out[it,i+1];
    v[i,it]=out[it,i+1+ng];
    u[i,it]=out[it,i+1+2*ng];
    p[i,it]=out[it,i+1+3*ng];
    m[i,it]=out[it,i+1+4*ng];
    t[it]=out[it,1];
  }
  }
```

- The solutions of eqs. (5.2) are displayed numerically in r and t with two fors.

```
#
# Display numerical solution
  iv=seq(from=1,to=nout,by=5);
  for(it in iv){
  cat(sprintf(
    "\n\n         t          r        n(r,t)"));
  cat(sprintf(
    "\n         t          r        v(r,t)"));
  cat(sprintf(
    "\n         t          r        u(r,t)"));
```

```
cat(sprintf(
  "\n        t            r        p(r,t)"));
cat(sprintf(
  "\n        t            r        m(r,t)"));
iv=seq(from=1,to=ng,by=10);
for(i in iv){
  cat(sprintf("\n%9.2e%11.2e%12.4f",
    t[it],r[i],n[i,it]));
  cat(sprintf("\n%9.2e%11.2e%12.4f",
    t[it],r[i],v[i,it]));
  cat(sprintf("\n%9.2e%11.2e%12.4f",
    t[it],r[i],u[i,it]));
  cat(sprintf("\n%9.2e%11.2e%12.4f",
    t[it],r[i],p[i,it]));
  cat(sprintf("\n%9.2e%11.2e%12.4f\n",
    t[it],r[i],m[i,it]));
  }
}
```

Every fifth value in *t* and every tenth value in *r* are displayed with `by=5,10`.

- The counter for the calls to `pde1a` is displayed at the end of the solution.

```
#
# Display ncall
  cat(sprintf("\n\n ncall = %2d",ncall));
```

- The solutions $n(r,t)$, $v(r,t)$, $u(r,t)$, $p(r,t)$, $m(r,t)$ are plotted in two dimensions (2D) with `matplot`.

```
#
# Plot numerical solutions
#
# 2D
  matplot(r,n,type="l",xlab="r",ylab="n(r,t)",
          lty=1,main="",lwd=2,col="black");
  matplot(r,v,type="l",xlab="r",ylab="v(r,t)",
          lty=1,main="",lwd=2,col="black");
  matplot(r,u,type="l",xlab="r",ylab="u(r,t)",
          lty=1,main="",lwd=2,col="black");
  matplot(r,p,type="l",xlab="r",ylab="p(r,t)",
          lty=1,main="",lwd=2,col="black");
  matplot(r,m,type="l",xlab="r",ylab="m(r,t)",
          lty=1,main="",lwd=2,col="black");
```

- The solutions $n(r,t)$, $v(r,t)$, $u(r,t)$, $p(r,t)$, $m(r,t)$ are plotted in three dimensions (3D) with `persp`.

```
  #
  # 3D
    persp(r,t,n,theta=45,phi=45,
          xlim=c(rl,ru),ylim=c(t0,tf),xlab="r",ylab="t",
          zlab="n(r,t)");
    persp(r,t,v,theta=45,phi=45,
          xlim=c(rl,ru),ylim=c(t0,tf),xlab="r",ylab="t",
          zlab="v(r,t)");
    persp(r,t,u,theta=45,phi=45,
          xlim=c(rl,ru),ylim=c(t0,tf),xlab="r",ylab="t",
          zlab="u(r,t)");
    persp(r,t,p,theta=60,phi=45,
          xlim=c(rl,ru),ylim=c(t0,tf),xlab="r",ylab="t",
          zlab="p(r,t)");
    persp(r,t,m,theta=60,phi=45,
          xlim=c(rl,ru),ylim=c(t0,tf),xlab="r",ylab="t",
          zlab="m(r,t)");
```

This completes the discussion of the main program in Listing 5.1. The ODE/MOL routine pde1a is considered next.

5.1.2   DODE routine

```
  pde1a=function(t,u5,parm){
#
# Function pde1a computes the t derivative
# vector of n(r,t), v(r,t), u(r,t), p(r,t),
# m(r,t)
#
# One vector to five vectors
  n=rep(0,ng);
  v=rep(0,ng);
  u=rep(0,ng);
  p=rep(0,ng);
  m=rep(0,ng);
  for(i in 1:ng){
    n[i]=u5[i];
    v[i]=u5[i+ng];
    u[i]=u5[i+2*ng];
    p[i]=u5[i+3*ng];
    m[i]=u5[i+4*ng];
  }
#
# Delayed variable vector
  if (t > tau){
    u5lag=lagvalue(t-tau);
  } else {
    u5lag=u0;
  }
```

```
#
# nr,vr,ur,pr,mr
  nr=dss004(rl,ru,ng,n);
  vr=dss004(rl,ru,ng,v);
  ur=dss004(rl,ru,ng,u);
  pr=dss004(rl,ru,ng,p);
  mr=dss004(rl,ru,ng,m);
#
# BCs
  nr[1]=kn*(n[1]-ncc);
  vr[1]=kv*(v[1]-vcc);
  ur[1]=ku*(u[1]-ucc);
  pr[1]=kp*(p[1]-pcc);
  mr[1]=km*(m[1]-mcc);
  nr[ng]=0;
  vr[ng]=0;
  ur[ng]=0;
  pr[ng]=0;
  mr[ng]=0;
#
# nrr,vrr,urr,prr,mrr
  nl=2;nu=2;
  nrr=dss044(rl,ru,ng,n,nr,nl,nu);
  vrr=dss044(rl,ru,ng,v,vr,nl,nu);
  urr=dss044(rl,ru,ng,u,ur,nl,nu);
  prr=dss044(rl,ru,ng,p,pr,nl,nu);
  mrr=dss044(rl,ru,ng,m,mr,nl,nu);
#
# n(r,t-tau)
  nd=rep(0,ng);
  for(i in 1:ng){
    nd[i]=u5lag[i];
  }
#
# PDEs
  nt=rep(0,ng); vt=rep(0,ng);
  ut=rep(0,ng); pt=rep(0,ng);
  mt=rep(0,ng);
  for(i in 1:ng){
    ri=2/r[i];
    nt[i]=Dn*(nrr[i]+ri*nr[i])-
          chiu*(n[i]*urr[i]+
                nr[i]*ur[i]+
                ri*n[i]*ur[i])-
          chip*(n[i]*prr[i]+
                nr[i]*pr[i]+
                ri*n[i]*pr[i])-
          chiv*(n[i]*vrr[i]+
                nr[i]*vr[i]+
```

```
                    ri*n[i]*vr[i])+
             mu1*n[i]*(1-n[i]);
       vt[i]=Dv*(vrr[i]+ri*vr[i])-
               delta*v[i]*m[i]+phi21*u[i]*p[i]-
               phi22*v[i]*p[i]+mu2*v[i]*(1-v[i]);
       ut[i]=Du*(urr[i]+ri*ur[i])-
               phi31*p[i]*u[i]-phi33*nd[i]*u[i]+
               alpha31*nd[i];
       pt[i]=Dp*(prr[i]+ri*pr[i])-
               phi41*p[i]*u[i]-phi42*p[i]*v[i]+
               alpha41*m[i];
       mt[i]=Dm*(mrr[i]+ri*mr[i])+
               phi52*p[i]*u[i]+phi53*nd[i]*u[i]-
               phi54*m[i];
   }
#
# Five vectors to one vector
  u5t=rep(0,(5*ng));
  for(i in 1:ng){
    u5t[i]      =nt[i];
    u5t[i+ng]   =vt[i];
    u5t[i+2*ng]=ut[i];
    u5t[i+3*ng]=pt[i];
    u5t[i+4*ng]=mt[i];
  }
#
# Increment calls to pde1a
  ncall<<-ncall+1;
#
# Return derivative vector
  return(list(c(u5t)));
}
```

Listing 5.2   DODE/MOL routine for eqs. (5.2), (5.3), (5.4) and (5.5).

We can note the following details about this listing.
- The function is defined.

```
    pde1a=function(t,u5,parm){
#
# Function pde1a computes the t derivative
# vector of n(r,t), v(r,t), u(r,t), p(r,t),
# m(r,t)
```

t is the current value of t in eqs. (5.2). u5 is the current numerical solution to eqs. (5.2). parm is an argument to pass parameters to pde1a (unused, but required in the argument list). The arguments must be listed in the order stated to properly interface with dede called in the main program of Listing 5.1. The DODE/MOL approximations of the derivatives

$\dfrac{\partial n(r,t)}{\partial t}, \dfrac{\partial v(r,t)}{\partial t}, \dfrac{\partial u(r,t)}{\partial t}, \dfrac{\partial p(r,t)}{\partial t}, \dfrac{\partial m(r,t)}{\partial t}$ of eqs. (5.2) are calculated and returned to dede as explained subsequently.

- The dependent variable vector, u5, is placed in five vectors to facilitate the programming of eqs. (5.2).

```
#
# One vector to five vectors
  n=rep(0,ng);
  v=rep(0,ng);
  u=rep(0,ng);
  p=rep(0,ng);
  m=rep(0,ng);
  for(i in 1:ng){
    n[i]=u5[i];
    v[i]=u5[i+ng];
    u[i]=u5[i+2*ng];
    p[i]=u5[i+3*ng];
    m[i]=u5[i+4*ng];
  }
```

- The five dependent variables in u5 (second input argument of pde1a) are lagged with lagvalue.

```
#
# Delayed variable vector
  if (t > tau){
    u5lag=lagvalue(t-tau);
  } else {
    u5lag=u0;
  }
```

u5 is a 255-vector with the five dependent variables of eqs. (5.2) placed according to the ICs programmed in the main program of Listing 5.1. u5lag has the lagged values of u5, so that $n(r, t - \tau)$ is placed in u5lag[1] to u5lag[51]. u0, the IC vector, is also the history vector when t <= tau.

- The first derivatives $\dfrac{\partial n(r,t)}{\partial r}, \dfrac{\partial v(r,t)}{\partial r}, \dfrac{\partial u(r,t)}{\partial r}, \dfrac{\partial p(r,t)}{\partial r}, \dfrac{\partial m(r,t)}{\partial r}$ are computed with dss004. The arguments of dss004 are explained in Appendix A1.

```
#
# nr,vr,ur,pr,mr
  nr=dss004(rl,ru,ng,n);
  vr=dss004(rl,ru,ng,v);
  ur=dss004(rl,ru,ng,u);
  pr=dss004(rl,ru,ng,p);
  mr=dss004(rl,ru,ng,m);
```

- Robin BCs (5.3) and Neumann BCs (5.4) are programmed.

```
#
# BCs
  nr[1]=kn*(n[1]-ncc);
  vr[1]=kv*(v[1]-vcc);
  ur[1]=ku*(u[1]-ucc);
  pr[1]=kp*(p[1]-pcc);
  mr[1]=km*(m[1]-mcc);
  nr[ng]=0;
  vr[ng]=0;
  ur[ng]=0;
  pr[ng]=0;
  mr[ng]=0;
```

Subscripts 1, nr correspond to $r = r_l, r_u$, respectively.

- The second derivatives $\dfrac{\partial^2 n(r,t)}{\partial r^2}$, $\dfrac{\partial^2 v(r,t)}{\partial r^2}$, $\dfrac{\partial^2 u(r,t)}{\partial r^2}$, $\dfrac{\partial^2 p(r,t)}{\partial r^2}$, $\dfrac{\partial^2 m(r,t)}{\partial r^2}$ in eqs. (5.2) are computed with dss044 (listed and discussed in Appendix A1).

```
#
# nrr,vrr,urr,prr,mrr
  nl=2;nu=2;
  nrr=dss044(rl,ru,ng,n,nr,nl,nu);
  vrr=dss044(rl,ru,ng,v,vr,nl,nu);
  urr=dss044(rl,ru,ng,u,ur,nl,nu);
  prr=dss044(rl,ru,ng,p,pr,nl,nu);
  mrr=dss044(rl,ru,ng,m,mr,nl,nu);
```

nl=nu=2 specify Neumann BCs for the second derivatives in r (BCs (5.3) are programmed as Neumann BCs, nl=2, since the first derivatives at $r = r_l$ are specified).

- The delayed variable $n(r, t - \tau)$ is extracted from u5lag as elements 1 to 51.

```
#
# n(r,t-tau)
  nd=rep(0,ng);
  for(i in 1:ng){
    nd[i]=u5lag[i];
  }
```

- The MOL programming of eqs. (5.2) steps through the 51 values of r in a for.

```
#
# DPDEs
  nt=rep(0,ng); vt=rep(0,ng);
```

```
ut=rep(0,ng); pt=rep(0,ng);
mt=rep(0,ng);
for(i in 1:ng){
  ri=2/r[i];
  nt[i]=Dn*(nrr[i]+ri*nr[i])-
        chiu*(n[i]*urr[i]+
              nr[i]*ur[i]+
              ri*n[i]*ur[i])-
        chip*(n[i]*prr[i]+
              nr[i]*pr[i]+
              ri*n[i]*pr[i])-
        chiv*(n[i]*vrr[i]+
              nr[i]*vr[i]+
              ri*n[i]*vr[i])+
        mu1*n[i]*(1-n[i]);
   vt[i]=Dv*(vrr[i]+ri*vr[i])-
         delta*v[i]*m[i]+phi21*u[i]*p[i]-
         phi22*v[i]*p[i]+mu2*v[i]*(1-v[i]);
   ut[i]=Du*(urr[i]+ri*ur[i])-
         phi31*p[i]*u[i]-phi33*nd[i]*u[i]+
         alpha31*nd[i];
   pt[i]=Dp*(prr[i]+ri*pr[i])-
         phi41*p[i]*u[i]-phi42*p[i]*v[i]+
         alpha41*m[i];
   mt[i]=Dm*(mrr[i]+ri*mr[i])+
         phi52*p[i]*u[i]+phi53*nd[i]*u[i]-
         phi54*m[i];
 }
```

Branching for $r = 0$ is not required since the interval in r is $1 \le r \le 10$ ($r = 0$ is not included). The correspondence of the DPDEs (eqs. (5.2)) and the programming indicates an important feature of the MOL.

- The five derivative vectors nt,vt,ut,pt,mt are placed in one derivative vector, u5t, to return to dede (called in the main program of Listing 5.1).

```
#
# Five vectors to one vector
  u5t=rep(0,(5*ng));
  for(i in 1:ng){
    u5t[i]      =nt[i];
    u5t[i+ng]   =vt[i];
    u5t[i+2*ng]=ut[i];
    u5t[i+3*ng]=pt[i];
    u5t[i+4*ng]=mt[i];
  }
```

- The counter for the calls to pde1a is incremented and returned to the main program by «-.

```
#
# Increment calls to pde1a
  ncall<<-ncall+1;
```

- The derivative vector u5t is returned to dede for the next step along the solution.

```
#
# Return derivative vector
  return(list(c(u5t)));
}
```

The derivative ut is returned as a list as required by dede. c is the R vector utility. The final } concludes pde1a.

The numerical and graphical output from the R routines of Listings 5.1 and 5.2 are considered next.

5.1.3 Numerical, graphical output

Abbreviated numerical output from Listings 5.1 and 5.2 follows in Table 5.4.

```
[1]  26

[1]  256
```

t	r	$n(r,t)$
t	r	$v(r,t)$
t	r	$u(r,t)$
t	r	$p(r,t)$
t	r	$m(r,t)$
0.00e+00	1.00e+00	0.0000
0.00e+00	1.00e+00	0.0000
0.00e+00	1.00e+00	0.0000
0.00e+00	1.00e+00	0.0000
0.00e+00	1.00e+00	0.0000
0.00e+00	2.80e+00	0.0000
0.00e+00	2.80e+00	0.0000
0.00e+00	2.80e+00	0.0000
0.00e+00	2.80e+00	0.0000
0.00e+00	2.80e+00	0.0000
0.00e+00	4.60e+00	0.0000
0.00e+00	4.60e+00	0.0000
0.00e+00	4.60e+00	0.0000
0.00e+00	4.60e+00	0.0000
0.00e+00	4.60e+00	0.0000

```
0.00e+00    6.40e+00       0.0000
0.00e+00    6.40e+00       0.0000
0.00e+00    6.40e+00       0.0000
0.00e+00    6.40e+00       0.0000
0.00e+00    6.40e+00       0.0000

0.00e+00    8.20e+00       0.0000
0.00e+00    8.20e+00       0.0000
0.00e+00    8.20e+00       0.0000
0.00e+00    8.20e+00       0.0000
0.00e+00    8.20e+00       0.0000

0.00e+00    1.00e+01       0.0000
0.00e+00    1.00e+01       0.0000
0.00e+00    1.00e+01       0.0000
0.00e+00    1.00e+01       0.0000
0.00e+00    1.00e+01       0.0000
               .              .
               .              .
               .              .
       Output from t=20,40,60,80
                removed
               .              .
               .              .
               .              .
           t          r        n(r,t)
           t          r        v(r,t)
           t          r        u(r,t)
           t          r        p(r,t)
           t          r        m(r,t)
1.00e+02    1.00e+00       0.7601
1.00e+02    1.00e+00       0.0516
1.00e+02    1.00e+00       0.4541
1.00e+02    1.00e+00       0.7620
1.00e+02    1.00e+00       0.5825

1.00e+02    2.80e+00       0.9843
1.00e+02    2.80e+00       0.0463
1.00e+02    2.80e+00       0.3140
1.00e+02    2.80e+00       0.5366
1.00e+02    2.80e+00       0.3167

1.00e+02    4.60e+00       0.9959
1.00e+02    4.60e+00       0.0503
1.00e+02    4.60e+00       0.3518
1.00e+02    4.60e+00       0.4268
1.00e+02    4.60e+00       0.2637
```

```
1.00e+02    6.40e+00      0.9968
1.00e+02    6.40e+00      0.0514
1.00e+02    6.40e+00      0.3744
1.00e+02    6.40e+00      0.3762
1.00e+02    6.40e+00      0.2445

1.00e+02    8.20e+00      0.9975
1.00e+02    8.20e+00      0.0517
1.00e+02    8.20e+00      0.3849
1.00e+02    8.20e+00      0.3553
1.00e+02    8.20e+00      0.2370

1.00e+02    1.00e+01      0.9977
1.00e+02    1.00e+01      0.0518
1.00e+02    1.00e+01      0.3878
1.00e+02    1.00e+01      0.3499
1.00e+02    1.00e+01      0.2351

ncall = 11977
```

Table 5.4   Abbreviated output from Listings 5.1 and 5.2

We can note the following details about this output.

- 26 output points in t as the first dimension of the solution matrix `uout` from `dede` as programmed in the main program of Listing 5.1.

- The solution matrix `uout` returned by `dede` has 256 elements as a second dimension. The first element is the value of t. Elements 2–256 in `uout` are $n(r, t)$, $v(r, t)$, $u(r, t)$, $p(r, t)$, $m(r, t)$ for eqs. (5.2).

- The solution is displayed for $t = 0, 20, ..., 100$ and $r = 1, 90/50, ..., 10$ as programmed in Listing 5.1 (every fifth value in t and every tenth value in r appear in Table 5.4).

- Homogeneous ICs (5.5) are confirmed (at $t = 0$).

- The computational effort is manageable, `ncall = 11977`, so that `dede` effectively computed a solution to eqs. (5.2).

The details of the solutions are presented in Figures 5.1-1,2,3,4,5 as 2D and in Figures 5.1-6,7,8,9,10 as 3D.

Figures 5.1 indicate the complexity of the solutions to eqs. (5.2) to (5.5). In particular,

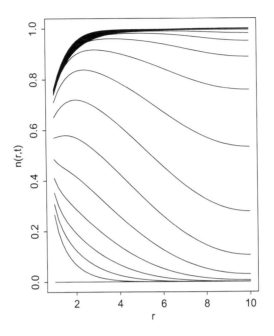

Figure 5.1-1   Numerical solution $n(r, t)$ from eq. (5.2-1), `matplot`.

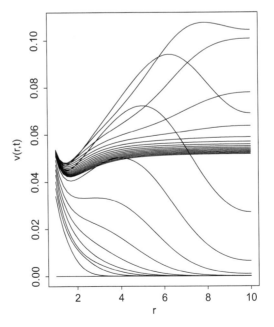

Figure 5.1-2   Numerical solution $v(r, t)$ from eq. (5.2-2), `matplot`.

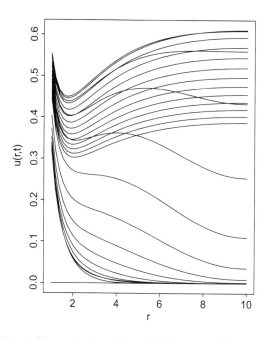

Figure 5.1-3   Numerical solution $u(r,t)$ from eq. (5.2-3), `matplot`.

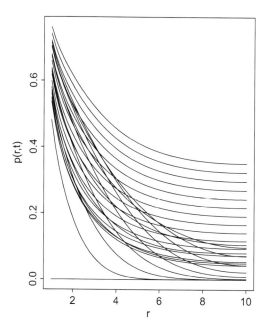

Figure 5.1-4   Numerical solution $p(r,t)$ from eq. (5.2-4), `matplot`.

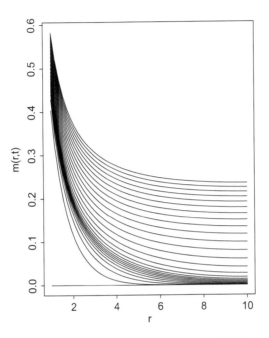

Figure 5.1-5   Numerical solution $m(r, t)$ from eq. (5.2-5), `matplot`.

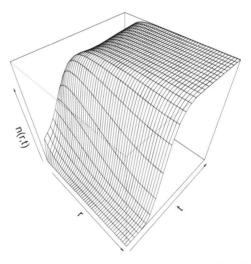

Figure 5.1-6   Numerical solution $n(r, t)$ from eq. (5.2-1), `persp`.

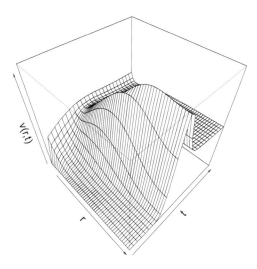

Figure 5.1-7   Numerical solution $v(r, t)$ from eq. (5.2-2), `persp`.

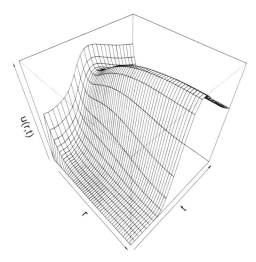

Figure 5.1-8   Numerical solution $u(r, t)$ from eq. (5.2-3), `persp`.

- Starting from the homogeneous ICs (eqs. (5.5)), the five dependent variables move through transients to a steady state (in Figures 5.1-1 to 5.1-5 note the movement from the IC to the steady state). This transition is stable and smooth.

- The metastasis gives a nearly uniform cancer cell density for large t (Figures 5.1-1,6). However, a somewhat counterintuitive condition in Figure 5.1-1 develops in which the slope of the solution curves near

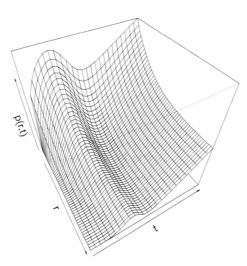

Figure 5.1-9   Numerical solution $p(r,t)$ from eq. (5.2-4), `persp`.

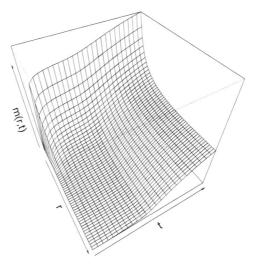

Figure 5.1-10   Numerical solution $m(r,t)$ from eq. (5.2-5), `persp`.

$r = r_l = 1$ changes from a negative to a positive value with increasing t. Since $n(r,t)$ is the only dependent variable with chemotaxis/haptotaxis terms, the change in slope might result from these terms.

To test this idea, if the coefficients in eq. (5.2-1) are changed to $\chi_u = \chi_p = \chi_v = 0$, the expected negative slope remains for $n(r,t)$ for large t as indicated in Figure 5.2-1. Also, $n(r,t)$ approaches a constant unit value for large t (rather than a variable $n(r,t)$ near $r = r_l = 1$ as in Figure 5.1-1).

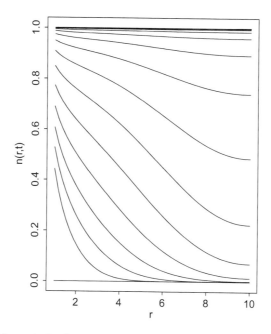

Figure 5.2-1 Numerical solution $n(r,t)$ from eq. (5.2-1), no chemotaxis/haptotaxis.

This result suggests that further consideration should be given to the use of chemotaxis/haptotaxis nonlinear diffusion in the balance for the cancer cell density (eq. (5.2-1)). This might be done by computing and displaying the individual RHS terms and the LHS derivative of eq. (5.2-1). This procedure is discussed in detail in [4, Chapter 4].

5.2 Summary and conclusions

The MOL analysis of the metastasis model, eqs. (5.2) to (5.5), is straightforward. For the model parameter values of Table 5.3, the numerical solutions of the DPDEs are stable and smooth. This contrasts with the solutions reported in [1, 2] which have sharp, moving peaks in r and t.

To study this possibility (sharp moving peaks) with the main program and DODE/MOL routine of Listings 5.1 and 5.2, ICs other than the homogeneous ICs used in this study would probably be required with a concomitant increase in the number of spatial points in r (from 51 to perhaps several thousands). However, for the model as started and implemented, the metastasis of cancer cells is clear (Figure 5.1-1), with perhaps further evaluation of the use of chemotaxis/haptotaxis (Figure 5.2-1).

The use of a delayed variable $(n(r, t - \tau))$ is a straightforward extension of the reported earlier models. The effect of the delay τ, and the coefficients χ_u, χ_p, χ_v, is left as an exercise for further study.

References

[1] Andasari, V., A. Gerisch, G. Lolas, A.P. South, and M.A.J. Chaplain (2011), Mathematical modeling of cancer cell invasion of tissue: Biological insight from mathematical analysis and computational simulation, *Journal of Mathematical Biology*, **63**, pp 141–171.

[2] Chaplain, M.A.J., and G. Lolas (2005), Mathematical modelling of cancer cell invasion of tissue: The role of the urokinase plasminogen activation system, *Mathematical Models and Methods in Applied Science*, **15**, no. 11, pp 1685–1734.

[3] Schiesser, W.E. (2016), *Method of Lines PDE Analysis in Biomedical Science and Engineering*, Wiley, Hoboken, NJ.

[4] Schiesser, W.E. (2019), *Numerical PDE Analysis of Retinal Neovascularization*, Elsevier, Cambridge, MA.

[5] Soetaert, K., J. Cash, and F. Mazzia (2012), *Solving Differential Equations in R*, Springer-Verlag, Heidelberg, Germany.

6

Subcutaneous Injection of Insulin

Introduction

In this chapter, a 3×3 (three equations in three unknowns) delay partial differential equation (DPDE) model is presented for the subcutaneous injection of insulin. The model dependent variables are the monomeric, dimeric, and hexameric forms of insulin. The spatiotemporal distribution of these forms of insulin in the subcutaneous tissue is computed by the method of lines (MOL) numerical integration of the DPDEs. The coding is in R as a main program and ordinary differential equation (ODE) ODE/MOL routine discussed subsequently.

6.1   DPDE model

The DPDE model is a modification of the model reported in [6]. The balances for the hexameric, dimeric, and monomeric insulin are

$$\frac{\partial h}{\partial t} = D\nabla^2 h + P(Qd^3 - h) + k_{hd}d(t - \tau) - k_h(h - h_a) \tag{6.1-1}$$

$$\frac{\partial d}{\partial t} = D\nabla^2 h - P(Qd^3 - h) - Bd + k_{dm}m(t - \tau) - 3k_{hd}d - k_d(d - d_a) \tag{6.1-2}$$

$$\frac{\partial m}{\partial t} = D\nabla^2 m - Bm - 2k_{dm}m - k_m(m - m_a) \tag{6.1-3}$$

The dependent variables of eqs. (6.1) are defined in Table 6.1.

Variable	Interpretation
h	hexameric insulin concentration
d	dimeric insulin concentration
$d(t - \tau)$	delayed d
m	monomeric insulin concentration
$m(t - \tau)$	delayed m
∇^2	Laplacian spatial differential operator

Table 6.1   Dependent variables of DPDEs (6.1)

Eqs. (6.1) are in a coordinate-free format. If they are expressed in one dimension (1D) spherical coordinates, that is, r in (r, θ, ϕ), they are

$$
\begin{aligned}
\frac{\partial h}{\partial t} &= D \frac{1}{r^2} \frac{\partial \left(r^2 \frac{\partial h}{\partial r} \right)}{\partial r} + P(Qd^3 - h) + k_{hd}d(r, t - \tau) - k_h(h - h_a) \\
&= D \left(\frac{\partial^2 h}{\partial r^2} + \frac{2}{r} \frac{\partial h}{\partial r} \right) + P(Qd^3 - h) + k_{hd}d(r, t - \tau) - k_h(h - h_a) \quad (6.2\text{-}1)
\end{aligned}
$$

$$
\begin{aligned}
\frac{\partial d}{\partial t} &= D \frac{1}{r^2} \frac{\partial \left(r^2 \frac{\partial d}{\partial r} \right)}{\partial r} - P(Qd^3 - h) - Bd + k_{dm}m(r, t - \tau) \\
&\quad - 3k_{hd}d - k_d(d - d_a) \\
&= D \left(\frac{\partial^2 d}{\partial r^2} + \frac{2}{r} \frac{\partial d}{\partial r} \right) - P(Qd^3 - h) - Bd + k_{dm}m(r, t - \tau) \\
&\quad - 3k_{hd}d - k_d(d - d_a) \quad (6.2\text{-}2)
\end{aligned}
$$

$$
\begin{aligned}
\frac{\partial m}{\partial t} &= D \frac{1}{r^2} \frac{\partial \left(r^2 \frac{\partial m}{\partial r} \right)}{\partial r} - Bm - 2k_{dm}m - k_m(m - m_a) \\
&= D \left(\frac{\partial^2 m}{\partial r^2} + \frac{2}{r} \frac{\partial m}{\partial r} \right) - Bm - 2k_{dm}m - k_m(m - m_a) \quad (6.2\text{-}3)
\end{aligned}
$$

Eqs. (6.2) are the starting point for the subsequent computer implementation and analysis.

To complete the model specification, eqs. (6.2) are second order in r and each requires two boundary conditions (BCs).

$$\frac{\partial h(r = r_l, t)}{\partial r} = \frac{\partial h(r = r_u, t)}{\partial r} = 0 \qquad\qquad (6.3\text{-}1,2)$$

$$\frac{\partial d(r = r_l, t)}{\partial r} = \frac{\partial d(r = r_u, t)}{\partial r} = 0 \qquad\qquad (6.3\text{-}3,4)$$

$$\frac{\partial m(r = r_l, t)}{\partial r} = \frac{\partial m(r = r_u, t)}{\partial r} = 0 \qquad\qquad (6.3\text{-}5,6)$$

Eqs. (6.3) are homogeneous Neumann BCs specifying symmetry in the solutions at $r = r_l = 0$ (eqs. (6.3-1,3,5)) and no flux at the outer boundary of the subcutaneous tissue at $r = r_u$ (eqs. (6.3-2,4,6)).

Eqs. (6.2) are first order in t and each requires one initial condition (IC).

$$h(r, t = 0) = h_0(r) \qquad\qquad (6.4\text{-}1)$$
$$d(r, t = 0) = d_0(r) \qquad\qquad (6.4\text{-}2)$$
$$m(r, t = 0) = m_0(r) \qquad\qquad (6.4\text{-}3)$$

where $h_0(r)$, $d_0(r)$, and $m_0(r)$ are functions to be specified.

The numerical values of the parameters (constants) in eqs. (6.2) are from Wach et al. [6] with parameters added as explained next.

Parameter	Value
Q	0.13
P	0.5
D	9.0×10^{-5}
B	1.3×10^{-2}
K_e	0.09
V_p	12
k_{hd}, k_{dm}	1.0×10^{-2}; 1.0×10^{-2}
τ	10
h_a, d_a, m_a	0; 0; 0
k_h, k_d, k_m	0; 1.0×10^{-3}; 0
k_{r1}, k_{r2}	1; 10

Table 6.2   Parameters of PDEs (6.2) (from Wach et al. [6, p. 20], with additions)

An interpretation of these numerical values is given in the following discussion of the terms of eqs. (6.2).

Eq. (6.2-1) is a balance on the hexameric insulin for an incremental volume $4\pi r^2 \Delta r$, followed by $\Delta r \to 0$, where r is the radial position in the subcutaneous tissue.

- $\dfrac{\partial h(r, t)}{\partial t}$: Increase (term > 0) or decrease (term < 0) of $h(r, t)$ with t.

- $D\left(\dfrac{\partial^2 h}{\partial r^2} + \dfrac{2}{r}\dfrac{\partial h}{\partial r}\right)$: Net rate of diffusion (random motion) of hexameric insulin into or out of the incremental volume.

- $+P(Qd^3 - h)$: Rate of increase of $h(r,t)$ from the dimeric and hexameric insulin. The forward rate is third order in $d(r,t)$, PQd^3, and the reverse rate is first order in $h(r,t)$, $-Ph$.

- $+k_{hd}d(r,t-\tau)$: Delayed rate of increase of $h(r,t)$ from $d(r,t-\tau)$.

- $-k_h(h - h_a)$: Rate of transfer of the hexameric insulin from the subcutaneous tissue to the bloodstream. k_h is a mass transfer coefficient and h_a is the insulin bloodstream concentration.

Eq. (6.2-2) is a balance on the dimeric insulin for an incremental volume $4\pi r^2 \Delta r$, followed by $\Delta r \to 0$, where r is the radial position in the subcutaneous tissue.

- $\dfrac{\partial d(r,t)}{\partial t}$: Increase (term > 0) or decrease (term < 0) of $d(r,t)$ with t.

- $D\left(\dfrac{\partial^2 d}{\partial r^2} + \dfrac{2}{r}\dfrac{\partial d}{\partial r}\right)$: Net rate of diffusion (random motion) of dimeric insulin into or out of the incremental volume.

- $-P(Qd^3 - h)$: Rate of decrease of $d(r,t)$ from the dimeric and hexameric insulin (the same term as in eq. (6.2-1) with a sign change). The forward rate is third order in $d(r,t)$, PQd^3, and the reverse rate is first order in $h(r,t)$, $-Ph$.

- $-Bd$: Decrease in $d(r,t)$ from bound dimeric insulin.

- $+k_{dm}m(r,t-\tau)$: Delayed rate of increase of $d(r,t)$ from $m(r,t-\tau)$.

- $-3k_{hd}d$: Rate of decrease of $d(r,t)$ from conversion to hexameric insulin.

- $-k_d(d - d_a)$: Rate of transfer of the dimeric insulin from the subcutaneous tissue to the bloodstream. k_d is a mass transfer coefficient and d_a is the insulin bloodstream concentration.

Eq. (6.2-3) is a balance on the monomeric insulin for an incremental volume $4\pi r^2 \Delta r$, followed by $\Delta r \to 0$, where r is the radial position in the subcutaneous tissue.

- $\dfrac{\partial m(r,t)}{\partial t}$: Increase (term > 0) or decrease (term < 0) of $m(r,t)$ with t.

- $D\left(\dfrac{\partial^2 m}{\partial r^2} + \dfrac{2}{r}\dfrac{\partial m}{\partial r}\right)$: Net rate of diffusion (random motion) of monomeric insulin into or out of the incremental volume.

- $-Bm$: Decrease in $m(r, t)$ from bound monomeric insulin.

- $-2k_{dm}m$: Rate of decrease of $m(r, t)$ from conversion to dimeric insulin.

- $-k_m(m - m_a)$: Rate of transfer of the monomeric insulin from the subcutaneous tissue to the bloodstream. k_m is a mass transfer coefficient and m_a is the insulin bloodstream concentration.

The diffusivity D is taken as a single value that can be used in eqs. (6.1). Different diffusivities for eqs. (6.1) with, for example, $D_h < D_d < D_m$ to reflect the different effective sizes for the hexameric, dimeric, and monomeric insulin molecules can easily be included in the model. This variation is left as an exercise.

This completes the specification of the DPDE model of eqs. (6.2), (6.3), (6.4). The computer implementation of the model is considered next.

6.1.1 Main program

The main program for eqs. (6.2), (6.3), (6.4) is in Listing 6.1.

```
#
# Three PDE model
#
# Delete previous workspaces
  rm(list=ls(all=TRUE))
#
# Access ODE integrator
  library("deSolve");
#
# Access functions for numerical solution
  setwd("f:/dpde/chap6");
  source("pde1a.R");
  source("dss004.R");
  source("dss044.R");
#
# Parameters
  Q=0.13;
  P=0.5;
  D=9.0e-05;
  B=1.3e-02;
  Ke=0.09;
  Vp=12;
  khd=1.0e-02; kdm=1.0e-02;
  tau=10;
  ha=0; da=0; ma=0;
  kh=0; kd=1.0e-03; km=0;
  kr1=1; kr2=10;
#
# Grid in r
```

```
  rl=0;ru=1;n=41;
  r=seq(from=rl,to=ru,by=(ru-rl)/(n-1));
#
# Initial condition
  u0=rep(0,(3*n));
  for(i in 1:n){
    u0[i]    =0;
    u0[i+n]  =0;
    u0[i+2*n]=kr1*exp(-kr2*r[i]^2);
  }
#
# Interval in t
  t0=0;tf=500;nout=26;
  tout=seq(from=t0,to=tf,by=(tf-t0)/(nout-1));
  ncall=0;
#
# ODE integration
  out=dede(y=u0,times=tout,func=pde1a);
  nrow(out);
  ncol(out);
#
# Store solution
  h=matrix(0,nrow=n,ncol=nout);
  d=matrix(0,nrow=n,ncol=nout);
  m=matrix(0,nrow=n,ncol=nout);
  t=rep(0,nout);
  for(it in 1:nout){
  for(i in 1:n){
    h[i,it]=out[it,i+1];
    d[i,it]=out[it,i+1+n];
    m[i,it]=out[it,i+1+2*n];
    t[it]=out[it,1];
  }
  }
#
# Display numerical solution
  iv=seq(from=1,to=nout,by=5);
  for(it in iv){
  cat(sprintf(
    "\n\n        t              r        h(r,t)"));
  cat(sprintf(
    "\n         t              r        d(r,t)"));
  cat(sprintf(
    "\n         t              r        m(r,t)"));
  iv=seq(from=1,to=n,by=5);
  for(i in iv){
    cat(sprintf("\n%9.2e%11.2e%12.4f",
      t[it],r[i],h[i,it]));
    cat(sprintf("\n%9.2e%11.2e%12.4f",
```

```
      t[it],r[i],d[i,it]));
    cat(sprintf("\n%9.2e%11.2e%12.4f\n",
      t[it],r[i],m[i,it]));
  }
  }
#
# Display ncall
  cat(sprintf("\n\n_ncall_=_%2d",ncall));
#
# Plot numerical solutions
#
# 2D
  matplot(r,h,type="l",xlab="r",ylab="h(r,t)",
          lty=1,main="",lwd=2,col="black");
  matplot(r,d,type="l",xlab="r",ylab="d(r,t)",
          lty=1,main="",lwd=2,col="black");
  matplot(r,m,type="l",xlab="r",ylab="m(r,t)",
          lty=1,main="",lwd=2,col="black");
#
# 3D
  persp(r,t,h,theta=45,phi=45,
        xlim=c(rl,ru),ylim=c(t0,tf),xlab="r",ylab="t",
        zlab="h(r,t)");
  persp(r,t,d,theta=45,phi=45,
        xlim=c(rl,ru),ylim=c(t0,tf),xlab="r",ylab="t",
        zlab="d(r,t)");
  persp(r,t,m,theta=45,phi=60,
        xlim=c(rl,ru),ylim=c(t0,tf),xlab="r",ylab="t",
        zlab="m(r,t)");
```

Listing 6.1   Main program for eqs. (6.2), (6.3), (6.4).

We can note the following details about Listing 6.1.

- Previous workspaces are deleted.

```
#
# Three PDE model
#
# Delete previous workspaces
  rm(list=ls(all=TRUE))
```

- The R ODE integrator library `deSolve` is accessed. Then the directory with the files for the solution of eqs. (6.2) is designated. Note that `setwd` (set working directory) uses / rather than the usual \.

```
#
# Access ODE integrator
  library("deSolve");
#
# Access functions for numerical solution
```

```
setwd("f:/dpde/chap6");
source("pde1a.R");
source("dss004.R");
source("dss044.R");
```

pde1a.R is the routine for eqs. (6.2) (discussed subsequently) based on the MOL, a general algorithm for partial differential equations [3]. dss004, dss044 are library routines for the calculation of first and second spatial derivatives. These routines are listed in Appendix A1 with additional explanation.

- The parameters of eqs. (6.2) as listed in Table 6.2 are programmed.

```
#
# Parameters
  Q=0.13;
  P=0.5;
  D=9.0e-05;
  B=1.3e-02;
  Ke=0.09;
  Vp=12;
  khd=1.0e-02; kdm=1.0e-02;
  tau=10;
  ha=0; da=0; ma=0;
  kh=0; kd=1.0e-03; km=0;
  kr1=1; kr2=10;
```

- A grid in r is defined for $r_l = 0 \le r \le r_u = 1$ with 41 points so $r = 0, 0.025, ..., 1$ cm. The value of ng was selected as large enough to avoid gridding effects in r, but small enough to avoid excessively large values of ncall.

```
#
# Grid in r
  rl=0;ru=1;n=41;
  r=seq(from=rl,to=ru,by=(ru-rl)/(n-1));
```

- An IC and history vector u0 is defined for $3*n=3*41=123$ points in r for eqs. (6.4).

```
#
# Initial condition
  u0=rep(0,(3*n));
  for(i in 1:n){
    u0[i]    =0;
    u0[i+n]  =0;
    u0[i+2*n]=kr1*exp(-kr2*r[i]^2);
  }
```

ICs (6.4-1,2) are homogeneous and IC (6.4-3) is a Gaussian function, centered at $r = r_l = 0$, which is a representation of the monomer insulin injection. The strength (volume) of the injection can be set with the selection of the constant k_{r1}.

- A temporal interval is defined with `nout=26` output points in t, initial and final values of `t0=0`, `tf=500`, so that $t = 0, 20, ..., 500$ (t is in min).

```
#
#  Interval in t
  t0=0;tf=500;nout=26;
  tout=seq(from=t0,to=tf,by=(tf-t0)/(nout-1));
  ncall=0;
```

The counter for the calls to `pde1a` is also initialized.

- The MOL/DODEs (delay ordinary differential equations) for eqs. (6.2) are integrated by the library integrator `dede` (available in `deSolve`, [5, Chapter 7]). As expected, the inputs to `dede` are the IC vector `u0`, the vector of output values of t, `times`, and the ODE function, `pde1a`. The length of `u0` (123) informs `dede` how many DODEs are to be integrated. `y`, `times`, `func` are reserved names.

```
#
#  ODE integration
#
#  ODE integration
  out=dede(y=u0,times=tout,func=pde1a);
  nrow(out);
  ncol(out);
```

- t is placed in vector `t` and $h(r,t)$, $d(r,t)$, $m(r,t)$ from eqs. (6.2) are placed in matrices `h,d,m` for numerical and graphical display.

```
#
#  Store solution
  h=matrix(0,nrow=n,ncol=nout);
  d=matrix(0,nrow=n,ncol=nout);
  m=matrix(0,nrow=n,ncol=nout);
  t=rep(0,nout);
  for(it in 1:nout){
  for(i in 1:n){
    h[i,it]=out[it,i+1];
    d[i,it]=out[it,i+1+n];
    m[i,it]=out[it,i+1+2*n];
    t[it]=out[it,1];
  }
  }
```

- The solutions of eqs. (6.2) are displayed numerically in r and t with two `for`s.

```
#
# Display numerical solution
  iv=seq(from=1,to=nout,by=5);
  for(it in iv){
  cat(sprintf(
    "\n\n          t              r        h(r,t)"));
  cat(sprintf(
    "\n          t              r        d(r,t)"));
  cat(sprintf(
    "\n          t              r        m(r,t)"));
  iv=seq(from=1,to=n,by=5);
  for(i in iv){
    cat(sprintf("\n%9.2e%11.2e%12.4f",
      t[it],r[i],h[i,it]));
    cat(sprintf("\n%9.2e%11.2e%12.4f",
      t[it],r[i],d[i,it]));
    cat(sprintf("\n%9.2e%11.2e%12.4f\n",
      t[it],r[i],m[i,it]));
  }
  }
```

Every fifth value in *t* and *r* is displayed with by=5.

- The counter for the calls to pde1a is displayed at the end of the solution.

```
#
# Display ncall
  cat(sprintf("\n\n ncall = %2d",ncall));
```

- The solutions $h(r,t)$, $d(r,t)$, $m(r,t)$ are plotted in two dimensions (2D) with matplot.

```
#
# Plot numerical solutions
#
# 2D
  matplot(r,h,type="l",xlab="r",ylab="h(r,t)",
          lty=1,main="",lwd=2,col="black");
  matplot(r,d,type="l",xlab="r",ylab="d(r,t)",
          lty=1,main="",lwd=2,col="black");
  matplot(r,m,type="l",xlab="r",ylab="m(r,t)",
          lty=1,main="",lwd=2,col="black");
```

- The solutions $h(r,t)$, $d(r,t)$, $m(r,t)$ are plotted in three dimensions (3D) with persp.

```
#
# 3D
  persp(r,t,h,theta=45,phi=45,
        xlim=c(rl,ru),ylim=c(t0,tf),xlab="r",ylab="t",
```

```
        zlab="h(r,t)");
  persp(r,t,d,theta=45,phi=45,
        xlim=c(rl,ru),ylim=c(t0,tf),xlab="r",ylab="t",
        zlab="d(r,t)");
  persp(r,t,m,theta=45,phi=60,
        xlim=c(rl,ru),ylim=c(t0,tf),xlab="r",ylab="t",
        zlab="m(r,t)");
```

This completes the discussion of the main program in Listing 6.1. The DODE/MOL routine pde1a is considered next.

6.1.2 DODE routine

```
  pde1a=function(t,u,parm){
#
# Function pde1a computes the t derivative
# vector of h(r,t), d(r,t), m(r,t)
#
# One vector to three vectors
  h=rep(0,n);
  d=rep(0,n);
  m=rep(0,n);
  for(i in 1:n){
    h[i]=u[i];
    d[i]=u[i+n];
    m[i]=u[i+2*n];
  }
#
# Delayed variable vector
  if (t > tau){
    ulag=lagvalue(t-tau);
  } else {
    ulag=u0;
  }
#
# hr,dr,mr
  hr=dss004(rl,ru,n,h);
  dr=dss004(rl,ru,n,d);
  mr=dss004(rl,ru,n,m);
#
# BCs
  hr[1]=0; hr[n]=0;
  dr[1]=0; dr[n]=0;
  mr[1]=0; mr[n]=0;
#
# hrr,drr,mrr
  nl=2;nu=2;
  hrr=dss044(rl,ru,n,h,hr,nl,nu);
```

```
  drr=dss044(rl,ru,n,d,dr,nl,nu);
  mrr=dss044(rl,ru,n,m,mr,nl,nu);
#
# h(r,t-tau),d(r,t-tau),m(r,t-tau)
  hd=rep(0,n);
  dd=rep(0,n);
  md=rep(0,n);
  for(i in 1:n){
    hd[i]=ulag[i];
    dd[i]=ulag[i+n];
    md[i]=ulag[i+2*n];
  }
#
# PDEs
  ht=rep(0,n); dt=rep(0,n);
  mt=rep(0,n);
  for(i in 1:n){
    rate=P*(Q*d[i]^3-h[i]);
    if(i==1){
      ht[i]=Dh*3*hrr[i]+rate+khd*dd[i]-kh*(h[i]-ha);
      dt[i]=Dd*3*drr[i]-rate-B*d[i]+kdm*md[i]-3*khd*d[i]-
            kd*(d[i]-da);
      mt[i]=Dm*3*mrr[i]-B*m[i]-2*kdm*m[i]-km*(m[i]-ma);
    }
    if(i>1){
    ri=2/r[i];
      ht[i]=Dh*(hrr[i]+ri*hr[i])+rate+khd*dd[i]-kh*(h[i]-ha);
      dt[i]=Dd*(drr[i]+ri*dr[i])-rate-B*d[i]+kdm*md[i]-
            3*khd*d[i]-kd*(d[i]-da);
      mt[i]=Dm*(mrr[i]+ri*mr[i])-B*m[i]-2*kdm*m[i]-
            km*(m[i]-ma);
    }
  }
#
# Three vectors to one vector
  ut=rep(0,(3*n));
  for(i in 1:n){
    ut[i]     =ht[i];
    ut[i+n]   =dt[i];
    ut[i+2*n]=mt[i];
  }
#
# Increment calls to pde1a
  ncall<<-ncall+1;
#
# Return derivative vector
  return(list(c(ut)));
}
```

Listing 6.2   DODE/MOL routine for eqs. (6.2), (6.3), (6.4).

We can note the following details about this listing.
- The function is defined.

```
pde1a=function(t,u,parm){
#
# Function pde1a computes the t derivative
# vector of h(r,t), d(r,t), m(r,t)
```

t is the current value of *t* in eqs. (6.2). u is the current numerical solution to eqs. (6.2). parm is an argument to pass parameters to pde1a (unused, but required in the argument list). The arguments must be listed in the order stated to properly interface with dede called in the main program of Listing 6.1. The DODE/MOL approximations of the derivatives $\dfrac{\partial h(r,t)}{\partial t}, \dfrac{\partial d(r,t)}{\partial t}, \dfrac{\partial m(r,t)}{\partial t}$ of eqs. (6.2) are calculated and returned to dede as explained subsequently.

- The dependent variable vector, u, is placed in three vectors to facilitate the programming of eqs. (6.2).

```
#
# One vector to three vectors
  h=rep(0,n);
  d=rep(0,n);
  m=rep(0,n);
  for(i in 1:n){
    h[i]=u[i];
    d[i]=u[i+n];
    m[i]=u[i+2*n];
  }
```

- The three dependent variables in u (second input argument of pde1a) are lagged with lagvalue.

```
#
# Delayed variable vector
  if (t > tau){
    ulag=lagvalue(t-tau);
  } else {
    ulag=u0;
  }
```

u is a 123-vector with the three dependent variables of eqs. (6.2) placed according to the ICs programmed in the main program of Listing 6.1. ulag has the lagged values of u. u0, the IC vector, is also the history vector when t <= tau.

- The first derivatives $\dfrac{\partial h(r,t)}{\partial r}, \dfrac{\partial d(r,t)}{\partial r}, \dfrac{\partial m(r,t)}{\partial r}$, are computed with dss004. The arguments of dss004 are explained in Appendix A1.

```
#
# hr,dr,mr
  hr=dss004(rl,ru,n,h);
  dr=dss004(rl,ru,n,d);
  mr=dss004(rl,ru,n,m);
```

- Homogeneous Neumann BCs (6.3) are programmed.

```
#
# BCs
  hr[1]=0; hr[n]=0;
  dr[1]=0; dr[n]=0;
  mr[1]=0; mr[n]=0;
```

Subscripts 1, n correspond to $r = r_l, r_u$, respectively.

- The second derivatives $\dfrac{\partial^2 h(r,t)}{\partial r^2}$, $\dfrac{\partial^2 d(r,t)}{\partial r^2}$, $\dfrac{\partial^2 m(r,t)}{\partial r^2}$ in eqs. (6.2) are computed with dss044 (listed and discussed in Appendix A1).

```
#
# hrr,drr,mrr
  nl=2;nu=2;
  hrr=dss044(rl,ru,n,h,hr,nl,nu);
  drr=dss044(rl,ru,n,d,dr,nl,nu);
  mrr=dss044(rl,ru,n,m,mr,nl,nu);
```

nl=nu=2 specify Neumann BCs for the second derivatives in *r* according to BCs (6.3).

- The lagged variables $h(r, t - \tau)$, $d(r, t - \tau)$, $m(r, t - \tau)$ are extracted from ulag.

```
#
# h(r,t-tau),d(r,t-tau),m(r,t-tau)
  hd=rep(0,n);
  dd=rep(0,n);
  md=rep(0,n);
  for(i in 1:n){
    hd[i]=ulag[i];
    dd[i]=ulag[i+n];
    md[i]=ulag[i+2*n];
  }
```

- The MOL programming of eqs. (6.2) steps through the $n = 41$ values of r in a for.

```
#
# PDEs
  ht=rep(0,n); dt=rep(0,n);
  mt=rep(0,n);
  for(i in 1:n){
```

```
    rate=P*(Q*d[i]^3-h[i]);
    if(i==1){
      ht[i]=Dh*3*hrr[i]+rate+khd*dd[i]-kh*(h[i]-ha);
      dt[i]=Dd*3*drr[i]-rate-B*d[i]+kdm*md[i]-3*khd*d[i]-
            kd*(d[i]-da);
      mt[i]=Dm*3*mrr[i]-B*m[i]-2*kdm*m[i]-km*(m[i]-ma);
    }
    if(i>1){
    ri=2/r[i];
      ht[i]=Dh*(hrr[i]+ri*hr[i])+rate+khd*dd[i]-kh*(h[i]-ha);
      dt[i]=Dd*(drr[i]+ri*dr[i])-rate-B*d[i]+kdm*md[i]-
            3*khd*d[i]-kd*(d[i]-da);
      mt[i]=Dm*(mrr[i]+ri*mr[i])-B*m[i]-2*kdm*m[i]-
            km*(m[i]-ma);
    }
  }
```

Branching for $r = 0$ (i=1) is required to accommodate the indeterminant forms $\dfrac{2}{r}\dfrac{\partial h}{\partial r} = 2\dfrac{\partial^2 h}{\partial r^2}, \dfrac{2}{r}\dfrac{\partial d}{\partial r} = 2\dfrac{\partial^2 d}{\partial r^2}, \dfrac{2}{r}\dfrac{\partial m}{\partial r} = 2\dfrac{\partial^2 m}{\partial r^2}$ at $r = 0$. The correspondence of the PDEs (eqs. (6.2)) and the programming indicates an important feature of the MOL.

- The three derivative vectors ht, dt, mt are placed in one derivative vector, ut, to return to dede (called in the main program of Listing 6.1).

```
#
# Three vectors to one vector
  ut=rep(0,(3*n));
  for(i in 1:n){
    ut[i]     =ht[i];
    ut[i+n]   =dt[i];
    ut[i+2*n]=mt[i];
  }
```

- The counter for the calls to pde1a is incremented and returned to the main program by «-.

```
#
# Increment calls to pde1a
  ncall<<-ncall+1;
```

- The derivative vector ut is returned to dede for the next step along the solution.

```
#
# Return derivative vector
  return(list(c(ut)));
}
```

The derivative ut is returned as a list as required by dede. c is the R vector utility. The final } concludes pde1a.

   The numerical and graphical output from the R routines of Listings 6.1
and 6.2 is considered next.

6.1.3   Numerical, graphical output

Abbreviated numerical output from Listings 6.1 and 6.2 follows in Table 6.3.

```
[1]  26

[1]  124
```

t	r	h(r,t)
t	r	d(r,t)
t	r	m(r,t)
0.00e+00	0.00e+00	0.0000
0.00e+00	0.00e+00	0.0000
0.00e+00	0.00e+00	1.0000
0.00e+00	1.25e-01	0.0000
0.00e+00	1.25e-01	0.0000
0.00e+00	1.25e-01	0.8553
0.00e+00	2.50e-01	0.0000
0.00e+00	2.50e-01	0.0000
0.00e+00	2.50e-01	0.5353
0.00e+00	3.75e-01	0.0000
0.00e+00	3.75e-01	0.0000
0.00e+00	3.75e-01	0.2451
0.00e+00	5.00e-01	0.0000
0.00e+00	5.00e-01	0.0000
0.00e+00	5.00e-01	0.0821
0.00e+00	6.25e-01	0.0000
0.00e+00	6.25e-01	0.0000
0.00e+00	6.25e-01	0.0201
0.00e+00	7.50e-01	0.0000
0.00e+00	7.50e-01	0.0000
0.00e+00	7.50e-01	0.0036
0.00e+00	8.75e-01	0.0000
0.00e+00	8.75e-01	0.0000
0.00e+00	8.75e-01	0.0005
0.00e+00	1.00e+00	0.0000
0.00e+00	1.00e+00	0.0000
0.00e+00	1.00e+00	0.0000

```
          .                    .
          .                    .
          .                    .
    Output for t=100,200,300,400
              removed

          .                    .
          .                    .
          .                    .
          t              r         h(r,t)
          t              r         d(r,t)
          t              r         m(r,t)
  5.00e+02    0.00e+00       0.0000
  5.00e+02    0.00e+00       0.0000
  5.00e+02    0.00e+00       0.0000

  5.00e+02    1.25e-01       0.0000
  5.00e+02    1.25e-01       0.0000
  5.00e+02    1.25e-01       0.0000

  5.00e+02    2.50e-01       0.0000
  5.00e+02    2.50e-01       0.0000
  5.00e+02    2.50e-01       0.0000

  5.00e+02    3.75e-01       0.0000
  5.00e+02    3.75e-01       0.0000
  5.00e+02    3.75e-01       0.0000

  5.00e+02    5.00e-01       0.0000
  5.00e+02    5.00e-01       0.0000
  5.00e+02    5.00e-01       0.0000

  5.00e+02    6.25e-01       0.0000
  5.00e+02    6.25e-01       0.0000
  5.00e+02    6.25e-01       0.0000

  5.00e+02    7.50e-01       0.0000
  5.00e+02    7.50e-01       0.0000
  5.00e+02    7.50e-01       0.0000

  5.00e+02    8.75e-01       0.0000
  5.00e+02    8.75e-01       0.0000
  5.00e+02    8.75e-01       0.0000

  5.00e+02    1.00e+00       0.0000
  5.00e+02    1.00e+00       0.0000
  5.00e+02    1.00e+00       0.0000

ncall = 3553
```

Table 6.3   Abbreviated output from Listings 6.1 and 6.2

We can note the following details about this output.

- 26 output points in t as the first dimension of the solution matrix uout from dede as programmed in the main program of Listing 6.1.

- The solution matrix uout returned by dede has 124 elements as a second dimension. The first element is the value of t. Elements 2–124 in uout are $h(r, t), d(r, t), m(r, t)$ for eqs. (6.2).

- The solution is displayed for $t = 0, 100, ..., 500$ and $r = 0, 0.125, ..., 1$ as programmed in Listing 6.1 (every fifth value in t and r appear in Table 6.3).

- Homogeneous ICs (6.4-1,2) and Gaussian IC (6.4-3) are confirmed (at $t = 0$).

- The computational effort is manageable, ncall = 3553, so that dede effectively computed a solution to eqs. (6.2).

The details of the solutions are presented in Figures 6.1-1–6.1-3 as 2D and in Figures 6.1-4–6.1-6 as 3D.

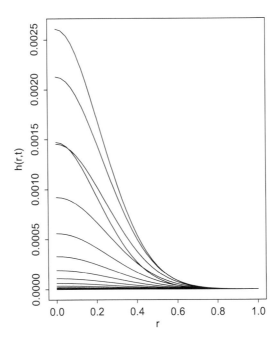

Figure 6.1-1   Numerical solution $h(r, t)$ from eq. (6.2-1), matplot.

Figures 6.1 indicate the complexity of the solutions to eqs. (6.2), (6.3), (6.4). In particular,

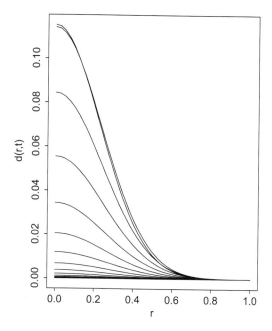

Figure 6.1-2 Numerical solution $d(r,t)$ from eq. (6.2-2), `matplot`.

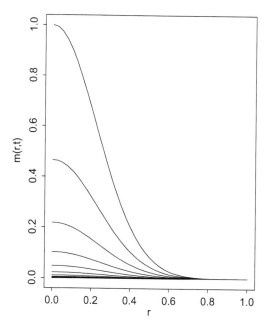

Figure 6.1-3 Numerical solution $m(r,t)$ from eq. (6.2-3), `matplot`.

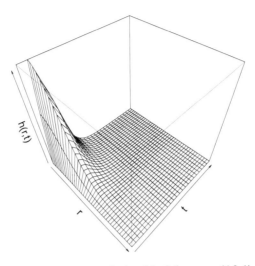

Figure 6.1-4   Numerical solution $h(r, t)$ from eq. (6.2-1), `persp`.

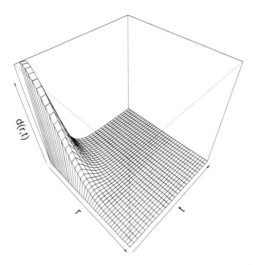

Figure 6.1-5   Numerical solution $d(r, t)$ from eq. (6.2-2), `persp`.

- Starting from the homogeneous ICs (eqs. (6.4-1,2)) and Gaussian IC (6.4-3), the three dependent variables move through transients to a zero steady state ($t = 500$ values in Table 6.3) as the effect of the Gaussian IC recedes with increasing t.

- The transients are a response to the Gaussian IC (6.4-3). If this IC is changed to zero, there is no source for moving the solutions away from the zero ICs which can be confirmed by changing the $m(r, t)$ IC in the main program of Listing 6.1. This is left as an exercise.

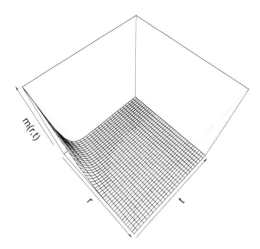

Figure 6.1-6 Numerical solution $m(r, t)$ from eq. (6.2-3), `persp`.

- The transfer of dimeric insulin to the bloodstream can be determined by computing and plotting $-k_d(d - d_a)$ from eq. (6.2-2) as a function of r and t. In particular, the effect of the mass transfer coefficient k_d on the flux of the dimeric insulin from the subcutaneous tissue to the bloodstream can be studied. This is left as an exercise (a detailed example of computing and displaying PDE terms is given in Chapter 4 [4]).

- The effect of the parameters of Table 6.2 on the DPDE solutions of eqs. (6.2) can be studied by changing the values of the parameters in the main program of Listing 6.1.

6.2 Summary and conclusions

The MOL solution of eqs. (6.2) is detailed in this chapter. In particular, the main program and DODE/MOL routines of Listings 6.1 and 6.2 can be used to study the DPDE model in detail, for example, the response to the Gaussian IC (6.4-3), variations in the parameters of Table 6.2, and changes in the form of the DPDEs.

As a word of caution, changes are probably possible that cause the routines to give excessively long execution times or fail to execute. Therefore, changes should be small and made incrementally. In this way, if the routines fail, the latest change is probably the source of the problem.

Wach et al. [6] is the starting point for the DPDE model of eqs. (6.2). Additional background information pertaining to subcutaneous insulin injection is given in [1, 2].

References

[1] Giang, D.V., Y. Lenbury, A.D. Gaetano, and P. Palumbo (2008), Delay model of glucose-insulin systems: Global stability and oscillated systems conditional on delays, *Journal Mathematical Analysis and Applications*, **343**, no. 2, pp 996–1006.

[2] Li, J., and J.D. Johnson (2009), Mathematical models of subcutaneous injection of insulin analogues: A mini-review, *Discrete and Continuous Dynamical Systems Series B*, **12**, no. 2, pp 401–414.

[3] Schiesser, W.E. (2016), *Method of Lines PDE Analysis in Biomedical Science and Engineering*, Wiley, Hoboken, NJ.

[4] Schiesser, W.E. (2019), *Numerical PDE Analysis of Retinal Neovascularization*, Elsevier, Cambridge, MA.

[5] Soetaert, K., J. Cash, and F. Mazzia (2012), *Solving Differential Equations in R*, Springer-Verlag, Heidelberg, Germany.

[6] Wach, P., Z. Trajanoski, P. Kotanko, and F. Skrabal (1995), Numerical approximation of mathematical model for absorption of subcutaneously injected insulin, *Medical and Biological Engineering and Computing*, **33**, no. 1, pp 18–23.

7

Hematopoietic Stem Cells

Introduction

This chapter pertains to the spatiotemporal distribution of hematopoietic stem cells (HSCs) which are the starting point for the various cells in the bloodstream. The following quotation gives a brief description of the functioning of HSCs [2]:

> A hematopoietic stem cell is a cell isolated from the blood or bone marrow that can renew itself, can differentiate to a variety of specialized cells, can mobilize out of the bone marrow into circulating blood, and can undergo programmed cell death, called apoptosis—a process by which cells that are detrimental or unneeded self-destruct.

A variant of the delayed partial differential equation (DPDE) model reported in [3] is coded in R as a main program and DODE/MOL routine in the method of lines (MOL) format [4].

7.1 DPDE model

The DPDE is

$$\frac{\partial u}{\partial t} = D\nabla^2 u - \delta u + \frac{\alpha u(t-\tau)}{1+\beta u^k(t-\tau)} \tag{7.1}$$

where Eq. (7.1) is in coordinate-free format. If a one-dimensional radial (polar) coordinate, r, is selected from cylindrical coordinates (r,θ,z), eq. (7.1) is

$$\frac{\partial u(r,t)}{\partial t} = D\left(\frac{\partial^2 u(r,t)}{\partial r^2} + \frac{1}{r}\frac{\partial u(r,t)}{\partial r}\right) - \delta u(r,t) + \frac{\alpha u(r,t-\tau)}{1+\beta u^k(r,t-\tau)} \tag{7.2}$$

Cylindrical coordinates were selected for a geometric (spatial) approximation of bone marrow.

Variable	Interpretation
u	HSC density
D	HSC diffusivity
∇^2	Laplacian spatial differential operator
$u(t-\tau)$	delayed u
τ	time delay between the initiation of the HSCs in the bone marrow and the release of mature HSCs into the bloodstream
δ	HSC death rate
α	intrinsic HSC growth rate
β	positive constant in the HSC growth rate function
k	positive integer ≥ 1

Table 7.1   Dependent variables of DPDE (7.1)

Eq. (7.2) is second order in r and requires two boundary conditions (BCs).

$$\frac{\partial u(r=r_l=0,t)}{\partial r}=0; \quad \frac{\partial u(r=r_u,t)}{\partial r}=k_u(u_a-u(r=r_u,t)) \qquad (7.3\text{-}1,2)$$

BC (7.3-1) is a symmetry (homogeneous Neumann) BC at the marrow center-line, $r=r_l=0$. BC (7.3-2) equates the HSC flux at the outer marrow boundary (proportional to $\frac{\partial u(r=r_u,t)}{\partial r}$) to the HSC flux into the bloodstream. This is a Robin BC since it includes both the dependent variable $u(r=r_u,t)$ and its first derivative $\frac{\partial u(r=r_u,t)}{\partial r}$. k_u is the ratio of a mass transfer coefficient to the diffusivity D.

Eq. (7.2) is first order in t and requires one initial condition (IC).

$$u(r,t=0)=u_0(r) \qquad (7.4)$$

where $u_0(r)$ is a function to be specified.

Eqs. (7.2), (7.3), (7.4) constitute the DPDE model. Routines for the numerical MOL solution follow.

7.1.1 Main program

A main program for eqs. (7.2), (7.3), (7.4) follows.

```
#
# One DPDE model
#
# Delete previous workspaces
  rm(list=ls(all=TRUE))
#
# Access ODE integrator
  library("deSolve");
#
# Access functions for numerical solution
  setwd("f:/dpde/chap7");
  source("pde1a.R");
  source("dss004.R");
  source("dss044.R");
#
# Parameters
  D=1.0e-03;
  tau=10;
  delta=1;
  k=1;
  alpha=5.0;
  beta=8.0;
  ku=1.0;
  ua=0;
  kr1=1; kr2=5;
#
# Grid in r
  rl=0;ru=1;n=101;
  r=seq(from=rl,to=ru,by=(ru-rl)/(n-1));
#
# Initial condition
  u0=rep(0,n);
  for(i in 1:n){
    u0[i]=kr1*exp(-kr2*r[i]^2);
  }
#
# Interval in t
  t0=0;tf=100;nout=26;
  tout=seq(from=t0,to=tf,by=(tf-t0)/(nout-1));
  ncall=0;
#
# ODE integration
  out=dede(y=u0,times=tout,func=pde1a);
  nrow(out);
  ncol(out);
#
```

```
# Store solution
  u=matrix(0,nrow=n,ncol=nout);
  t=rep(0,nout);
  for(it in 1:nout){
  for(i in 1:n){
    u[i,it]=out[it,i+1];
    t[it]=out[it,1];
  }
  }
#
# Display numerical solution
  iv=seq(from=1,to=nout,by=5);
  for(it in iv){
  cat(sprintf(
    "\n\n          t                r         u(r,t)"));
  iv=seq(from=1,to=n,by=10);
  for(i in iv){
    cat(sprintf("\n%9.2e%11.2e%12.4f",
      t[it],r[i],u[i,it]));
  }
  }
#
# Display ncall
  cat(sprintf("\n\n ncall = %2d",ncall));
#
# Plot numerical solutions
#
# 2D
  matplot(r,u,type="l",xlab="r",ylab="u(r,t)",
          lty=1,main="",lwd=2,col="black");
#
# 3D
  persp(r,t,u,theta=45,phi=45,
        xlim=c(rl,ru),ylim=c(t0,tf),xlab="r",
        ylab="t",zlab="u(r,t)");
```

Listing 7.1   Main program for eqs. (7.2), (7.3), (7.4).

We can note the following details about Listing 7.1.

- Previous workspaces are deleted.

```
  #
  # One DPDE model
  #
  # Delete previous workspaces
    rm(list=ls(all=TRUE))
```

- The R ODE integrator library deSolve is accessed. Then the directory with the files for the solution of eq. (7.2) is designated. Note that setwd (set working directory) uses / rather than the usual \.

```
#
# Access ODE integrator
  library("deSolve");
#
# Access functions for numerical solution
  setwd("f:/dpde/chap7");
  source("pde1a.R");
  source("dss004.R");
  source("dss044.R");
```

pde1a.R is the routine for eq. (7.2) (discussed subsequently) based on the MOL, a general algorithm for partial differential equations (PDEs) [4]. dss004, dss044 are library routines for the calculation of first and second spatial derivatives. These routines are listed in Appendix A1 with additional explanation.

- The parameters of eqs. (7.2), (7.3-2) as listed in Table 7.1 are programmed.

```
#
# Parameters
  D=1.0e-03;
  tau=10;
  delta=1;
  k=1;
  alpha=5.0;
  beta=8.0;
  ku=1.0;
  ua=0;
  kr1=1; kr2=5;
```

These values are variants of the parameters in [3] with additions as required by eqs. (7.2), (7.3), (7.4).

- A grid in r is defined for $r_l = 0 \leq r \leq r_u = 1$ cm with 101 points so $r = 0, 0.1, ..., 1$.

```
#
# Grid in r
  rl=0;ru=1;n=101;
  r=seq(from=rl,to=ru,by=(ru-rl)/(n-1));
```

- An IC and history vector u0 is defined for n=101 points in r for eq. (7.4).

```
#
# Initial condition
  u0=rep(0,n);
  for(i in 1:n){
    u0[i]=kr1*exp(-kr2*r[i]^2);
  }
```

IC (7.4) is a Gaussian function, centered at $r = r_l = 0$, which is a representation of the initial HSC distribution. The solution of the DPDE model, eqs. (7.2), (7.3), (7.4), gives the evolution of $u(r,t)$ away from this IC.

- A temporal interval is defined with nout=26 output points in t, initial and final values of t0=0, tf=100, so that $t = 0, 4, ..., 100$.

```
#
# Interval in t
  t0=0;tf=100;nout=26;
  tout=seq(from=t0,to=tf,by=(tf-t0)/(nout-1));
  ncall=0;
```

If space and time are in cm and h, the diffusivity is (Listing 7.1) $D = 10^{-3} \text{cm}^2/\text{h} = 2.78 \times 10^{-7} \text{ cm}^2/\text{s}$ which is a representative (nominal physical) value for a liquid diffusivity.

The counter for the calls to pde1a is also initialized.

- The MOL/DODEs for eq. (7.2) are integrated by the library integrator dede (available in deSolve, [6, Chapter 7]). As expected, the inputs to dede are the IC vector u0, the vector of output values of t, times, and the ODE function, pde1a. The length of u0 (101) informs dede how many DODEs are to be integrated. y, times, func are reserved names.

```
#
# ODE integration
  out=dede(y=u0,times=tout,func=pde1a);
  nrow(out);
  ncol(out);
```

- t is placed in vector t and $u(r,t)$ from eq. (7.2) is placed in matrix u for numerical and graphical display.

```
#
# Store solution
  u=matrix(0,nrow=n,ncol=nout);
  t=rep(0,nout);
  for(it in 1:nout){
  for(i in 1:n){
    u[i,it]=out[it,i+1];
    t[it]=out[it,1];
  }
  }
```

- The solution of eq. (7.2) is displayed numerically in r and t with two fors.

```
#
# Display numerical solution
  iv=seq(from=1,to=nout,by=5);
  for(it in iv){
```

```
cat(sprintf(
  "\n\n          t          r        u(r,t)"));
iv=seq(from=1,to=n,by=10);
for(i in iv){
  cat(sprintf("\n%9.2e%11.2e%12.4f",
    t[it],r[i],u[i,it]));
}
}
```

Every fifth value in t and every tenth value in r are displayed with by=5,10.

- The counter for the calls to pde1a is displayed at the end of the solution.

```
#
# Display ncall
  cat(sprintf("\n\n ncall = %2d",ncall));
```

- The solution $u(r,t)$ is plotted in two dimensions (2D) with matplot.

```
#
# Plot numerical solutions
#
# 2D
  matplot(r,u,type="l",xlab="r",ylab="u(r,t)",
          lty=1,main="",lwd=2,col="black");
```

- The solutions $u(r,t)$ are plotted in three dimensions (3D) with persp.

```
#
# 3D
  persp(r,t,u,theta=45,phi=45,
        xlim=c(rl,ru),ylim=c(t0,tf),xlab="r",
        ylab="t",zlab="u(r,t)");
```

This completes the discussion of the main program in Listing 7.1. The DODE/MOL routine pde1a is considered next.

7.1.2 DODE routine

Routine pde1a is in Listing 7.2.

```
pde1a=function(t,u,parm){
#
# Function pde1a computes the t derivative
# vector of u(r,t)
#
# Delayed variable vector
  if (t > tau){
    ulag=lagvalue(t-tau);
  } else {
```

```
    ulag=u0;
  }
#
# ur
  ur=dss004(rl,ru,n,u);
#
# BCs
  ur[1]=0; ur[n]=ku*(ua-u[n]);
#
# urr
  nl=2;nu=2;
  urr=dss044(rl,ru,n,u,ur,nl,nu);
#
# DPDEs
  ut=rep(0,n);
  for(i in 1:n){
    rate=alpha*ulag[i]/(1+beta*ulag[i]^k);
    if(i==1){
      ut[i]=D*2*urr[i]-delta*u[i]+rate;
    }
    if(i>1){
    ri=1/r[i];
      ut[i]=D*(urr[i]+ri*ur[i])-delta*u[i]+rate;
    }
  }
#
# Increment calls to pde1a
  ncall<<-ncall+1;
#
# Return derivative vector
  return(list(c(ut)));
}
```

Listing 7.2   DODE/MOL routine for eqs. (7.2), (7.3), (7.4).

We can note the following details about this listing.

- The function is defined.

```
    pde1a=function(t,u,parm){
#
# Function pde1a computes the t derivative
# vector of u(r,t)
```

t is the current value of t in eq. (7.2). u is the current numerical solution to eq. (7.2). parm is an argument to pass parameters to pde1a (unused, but required in the argument list). The arguments must be listed in the order stated to properly interface with dede called in the main program of Listing 7.1. The DODE/MOL approximation of the derivative $\dfrac{\partial u(r,t)}{\partial t}$ of eq. (7.2) is calculated and returned to dede as explained subsequently.

- The dependent variable in u (second input argument of pde1a) is lagged with lagvalue.

```
#
# Delayed variable vector
  if (t > tau){
    ulag=lagvalue(t-tau);
  } else {
    ulag=u0;
  }
```

u is a 101-vector with the dependent variable of eq. (7.2) placed according to the IC programmed in the main program of Listing 7.1. ulag has the lagged values of u. u0, the IC vector, is also the history vector when t <= tau.

- The first derivative $\dfrac{\partial u(r,t)}{\partial r}$ is computed with dss004. The arguments of dss004 are explained in Appendix A1.

```
#
# ur
  ur=dss004(rl,ru,n,u);
```

- Homogeneous Neumann BC (7.3-1) and Robin BC (7.3-2) are programmed.

```
#
# BCs
  ur[1]=0; ur[n]=ku*(ua-u[n]);
```

Subscripts 1, n correspond to $r = r_l, r_u$, respectively.

- The second derivative $\dfrac{\partial^2 u(r,t)}{\partial r^2}$ in eq. (7.2) is computed with dss044 (listed and discussed in Appendix A1).

```
#
# urr
  nl=2;nu=2;
  urr=dss044(rl,ru,n,u,ur,nl,nu);
```

nl=nu=2 specify Neumann and Robin BCs for the second derivative in r according to BCs (7.3).

- The MOL programming of eq. (7.2) steps through the $n = 101$ values of r in a for.

```
#
# DPDEs
  ut=rep(0,n);
  for(i in 1:n){
```

```
        rate=alpha*ulag[i]/(1+beta*ulag[i]^k);
        if(i==1){
          ut[i]=D*2*urr[i]-delta*u[i]+rate;
        }
        if(i>1){
        ri=1/r[i];
          ut[i]=D*(urr[i]+ri*ur[i])-delta*u[i]+rate;
        }
      }
```

Branching for $r = 0$ (i=1) is required to accommodate the indeterminant form $\dfrac{1}{r}\dfrac{\partial u}{\partial r} = \dfrac{\partial^2 u}{\partial r^2}$ at $r = 0$. The correspondence of the PDE (eq. (7.2)) and the programming indicates an important feature of the MOL.

- The counter for the calls to pde1a is incremented and returned to the main program by <<-.

```
#
# Increment calls to pde1a
  ncall<<-ncall+1;
```

- The derivative vector ut is returned to dede for the next step along the solution.

```
#
# Return derivative vector
  return(list(c(ut)));
}
```

The derivative ut is returned as a list as required by dede. c is the R vector utility. The final } concludes pde1a.

The numerical and graphical output from the R routines of Listings 7.1 and 7.2 is considered next.

7.1.3 Numerical, graphical output

Abbreviated numerical output from Listings 7.1 and 7.2 follows in Table 7.2.

```
[1]  26

[1]  102
             t                 r          u(r,t)
  0.00e+00      0.00e+00          1.0000
  0.00e+00      1.00e-01          0.9512
  0.00e+00      2.00e-01          0.8187
  0.00e+00      3.00e-01          0.6376
  0.00e+00      4.00e-01          0.4493
```

```
0.00e+00    5.00e-01      0.2865
0.00e+00    6.00e-01      0.1653
0.00e+00    7.00e-01      0.0863
0.00e+00    8.00e-01      0.0408
0.00e+00    9.00e-01      0.0174
0.00e+00    1.00e+00      0.0067
   .                         .
   .                         .
   .                         .

     Output for t=20,40,60,80
               removed

   .                         .
   .                         .
   .                         .

        t           r       u(r,t)
1.00e+02    0.00e+00      0.5000
1.00e+02    1.00e-01      0.5000
1.00e+02    2.00e-01      0.5000
1.00e+02    3.00e-01      0.5000
1.00e+02    4.00e-01      0.5000
1.00e+02    5.00e-01      0.5000
1.00e+02    6.00e-01      0.5000
1.00e+02    7.00e-01      0.5000
1.00e+02    8.00e-01      0.4999
1.00e+02    9.00e-01      0.4989
1.00e+02    1.00e+00      0.4826

ncall = 9267
```

Table 7.2 Abbreviated output from Listings 7.1 and 7.2

We can note the following details about this output.

- 26 output points in t as the first dimension of the solution matrix uout from dede as programmed in the main program of Listing 7.1.

- The solution matrix uout returned by dede has 102 elements as a second dimension. The first element is the value of t. Elements 2–102 in uout are $u(r, t)$ for eq. (7.2).

- The solution is displayed for $t = 0, 20, ..., 100$ and $r = 1, 0.1, ..., 1$ as programmed in Listing 7.1 (every fifth value in t and every tenth value r appear in Table 7.2).

- Gaussian IC (7.4) is confirmed (at $t = 0$).

- The computational effort is manageable, ncall $= 9267$, so that dede effectively computed a solution to eq. (7.2).

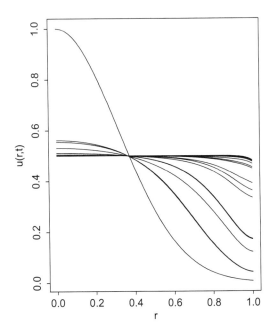

Figure 7.1-1   Numerical solution $u(r,t)$ from eq. (7.2), `matplot`.

The details of the solutions are presented in Figure 7.1-1 as 2D and in Figure 7.1-2 as 3D.

Figures 7.1 indicate that the Gaussian IC (7.4) forms a front moving left to right. The solution curves are not equally spaced in t which probably results from the nonlinearity of the generation term $\dfrac{\alpha u(r,t-\tau)}{1+\beta u^k(r,t-\tau)}$

The solution approaches a steady state $u(r,t \to \infty) = 0.5$ with no variation in r. This can be confirmed by numerical substitution in the RHS of eq. (7.2).

$$D\left(\frac{\partial^2 u(r,t)}{\partial r^2} + \frac{1}{r}\frac{\partial u(r,t)}{\partial r}\right) - \delta u(r,t) + \frac{\alpha u(r,t-\tau)}{1+\beta u^k(r,t-\tau)}$$

$$= 1.0 \times 10^{-3}\left(0 + \frac{1}{r}(0)\right) - (1)(0.5) + \frac{(5)(0.5)}{1+(8)(0.5)^1} = 0$$

so that from eq. (7.2) $\dfrac{\partial u(r,t \to \infty)}{\partial t} = 0$ (stable steady state).

The effect of the parameters of Table 7.1 on the solutions of eq. (7.2) can be studied by changing the values of the parameters in the main program of Listing 7.1.

For example, if $\alpha = 0$ in eq. (7.2), there is no generation of HSCs and the steady state is $u(r,t \to \infty) = 0$, that is, all of the HSCs enter the bloodstream through BC (7.3-2). Confirmation of this result is left as an exercise.

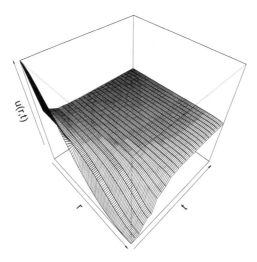

Figure 7.1-2 Numerical solution $u(r,t)$ from eq. (7.2), `persp`.

The effect of the parameter k in the source term can be studied by changing the value $k = 1$ in Listing 7.1. For example, if $k = 2$, the abbreviated numerical output is in Table 7.3.

```
[1] 26

[1] 102
```

```
        t             r          u(r,t)
0.00e+00      0.00e+00          1.0000
0.00e+00      1.00e-01          0.9512
0.00e+00      2.00e-01          0.8187
0.00e+00      3.00e-01          0.6376
0.00e+00      4.00e-01          0.4493
0.00e+00      5.00e-01          0.2865
0.00e+00      6.00e-01          0.1653
0.00e+00      7.00e-01          0.0863
0.00e+00      8.00e-01          0.0408
0.00e+00      9.00e-01          0.0174
0.00e+00      1.00e+00          0.0067

      .             .
      .             .
      .             .

   Output for t=20,40,60,80
           removed

      .             .
      .             .
      .             .

        t             r          u(r,t)
1.00e+02      0.00e+00          0.7071
```

```
1.00e+02    1.00e-01     0.7071
1.00e+02    2.00e-01     0.7072
1.00e+02    3.00e-01     0.7072
1.00e+02    4.00e-01     0.7072
1.00e+02    5.00e-01     0.7073
1.00e+02    6.00e-01     0.7074
1.00e+02    7.00e-01     0.7072
1.00e+02    8.00e-01     0.7068
1.00e+02    9.00e-01     0.7063
1.00e+02    1.00e+00     0.6892

ncall = 13356
```

Table 7.3   Abbreviated output from Listings 7.1 and 7.2, $k = 2, n = 101$

We can note the following details about this output.

- 26 output points in t as the first dimension of the solution matrix uout from dede as programmed in the main program of Listing 7.1.

- The solution matrix uout returned by dede has 102 elements as a second dimension. The first element is the value of t. Elements 2–102 in uout are $u(r,t)$ for eq. (7.2).

- The solution is displayed for $t = 0, 20, ..., 100$ and $r = 1, 0.1, ..., 2$ as programmed in Listing 7.1 (every fifth value in t and every tenth value of r appear in Table 7.3).

- Gaussian IC (7.4) is confirmed (at $t = 0$).

- The computational effort is ncall = 13356 is manageable which reflects the effectiveness of dede.

The details of the solutions are presented in Figure 7.2-1 as 2D and in Figure 7.2-2 as 3D.

Figures 7.2 indicate an oscillatory solution before approaching a steady state. Resolution of the complicated transient is the reason for the selection of the relatively fine grid in r with n=101. This oscillatory effect is also reported in [3], Figures (A.2), (A.3) and [1] resulting from the nonlinear source term $\dfrac{\alpha u(r, t - \tau)}{1 + \beta u^k(r, t - \tau)}$ (which includes k).

The solution approaches a steady state $u(r, t \to \infty) = 0.7071$ with no variation in r with increasing t. This can be confirmed by numerical substitution in the RHS of eq. (7.2).

$$D\left(\frac{\partial^2 u(r,t)}{\partial r^2} + \frac{1}{r}\frac{\partial u(r,t)}{\partial r}\right) - \delta u(r,t) + \frac{\alpha u(r, t - \tau)}{1 + \beta u^k(r, t - \tau)}$$

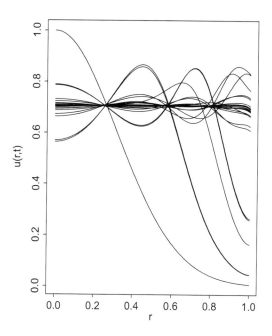

Figure 7.2-1   Numerical solution $u(r,t)$ from eq. (7.2), $k = 2$, `matplot`.

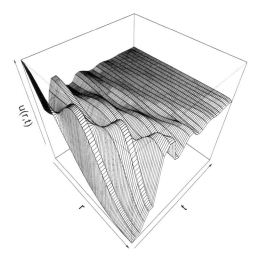

Figure 7.2-2   Numerical solution $u(r,t)$ from eq. (7.2), $k = 2$, `persp`.

$$= 1.0 \times 10^{-3} \left(0 + \frac{1}{r}(0) \right) - (1)(0.7071) + \frac{(5)(0.7071)}{1 + (8)(0.7071)^2} = 1.085 \times 10^{-5} \approx 0$$

so that from eq. (7.2) $\dfrac{\partial u(r, t \to \infty)}{\partial t} \approx 0$ (stable steady state).

These cases ($k = 1, 2$) demonstrate that k is a sensitive parameter that can determine the qualitative form of the solution of eqs. (7.2), (7.3), (7.4) (cf. Figures 7.1 and 7.2) through the nonlinear source term $\dfrac{\alpha u(r, t - \tau)}{1 + \beta u^k(r, t - \tau)}$. In this way, α and τ could also be sensitive parameters. Further study, which could include variation of these parameters with t to represent the development of a blood-borne disease, is left as an exercise.

7.2 Summary and conclusions

The numerical MOL integration of the DPDE model of eqs. (7.2), (7.3), (7.4) is straightforward. k in the HSC nonlinear source term of eq. (7.2) is an important parameter in determining the form of the spatiotemporal distribution of the HSC density. Numerical experimentation with the model can be used to identify other properties of the solution. For example, the flux from BC (7.3-2), $Dk_u(u_a - u(r = r_u, t))$, can easily be computed from the solution $u(r, t)$ to determine the rate of transfer of the HSCs from the bone marrow to the bloodstream. The computation and display of the RHS and LHS terms of eq. (7.2) from the solution $u(r, t)$ can give further insight into the source of the solution features. A detailed example of this procedure is given in [5, Chapter 4].

References

[1] Mackey, M.C., C. Ou, L. Pujo-Menjouet, and J. Wu (2006), Periodic oscillations of blood cell populations in chronic myelogenous leukemia, *SIAM Journal Mathematical Analysis*, **38**, no. 1, pp 166–187.

[2] National Institutes of Health, Stem Cell Information, Chapter 5, Hematopoietic Stem Cells.

[3] Pan, X., H. Shu, L. Wang, and X.-S. Wang (2019), Dirichlet problem for a delayed diffusive hematopoiesis model, *Nonlinear Analysis: Real World Problems*, **48**, pp 493–516.

[4] Schiesser, W.E. (2016), *Method of Lines PDE Analysis in Biomedical Science and Engineering*, Wiley, Hoboken, NJ.

[5] Schiesser, W.E. (2019), *Numerical PDE Analysis of Retinal Neovascularization*, Elsevier, Cambridge, MA.

[6] Soetaert, K., J. Cash, and F. Mazzia (2012), *Solving Differential Equations in R*, Springer-Verlag, Heidelberg, Germany.

8

Drug Eluting Stents

Introduction

Reduced arterial blood flow, often from the accumulation of plaque (atherosclerosis), is a major health concern that is treated by surgery (by-pass) or by inserting a drug eluting stent where the blockage of blood flow has occurred. The stent widens the artery and thus achieves increased blood flow.

A stent is a tube-like hollow metallic cylinder. A principal concern for the continued performance of the stent is blood clotting (coagulation) around the stent. To reduce this dangerous condition, the stent is coated on the external surface, adjacent to the arterial wall, with a drug that is an anticoagulant (antiplatelet).

In the first example of this chapter, a three DPDE model is formulated to represent the stent drug coating and the adjacent arterial wall. The spatial domain is modeled as two concentric cylinders. The inner cylinder is the coating on the stent external surface that contains the drug. The outer concentric cylinder in the arterial wall that receives the drug from the stent. The statement of the model is considered next, followed by a method of lines (MOL) implementation in R [1].

8.1   DPDE model, nondegradable stent coating

The following three DPDE models are variants of the model reported in [4].

$$\frac{\partial C_1(r,t)}{\partial t} = D_1 \left(\frac{\partial^2 C_1(r,t)}{\partial r^2} + \frac{1}{r}\frac{\partial C_1(r,t)}{\partial r} \right); \ r_{l1} \le r \le r_{u1} \qquad (8.1\text{-}1)$$

$$\frac{\partial C_2(r,t)}{\partial t} = D_2 \left(\frac{\partial^2 C_2(r,t)}{\partial r^2} + \frac{1}{r}\frac{\partial C_2(r,t)}{\partial r} \right)$$

$$- k_a(S_0 - B_2(r,t))C_2(r,t-\tau) + k_d B_2(r,t); \ r_{l2} \le r \le r_{u2} \qquad (8.1\text{-}2)$$

$$\frac{\partial B_2(r,t)}{\partial t} = k_a(S_0 - B_2(r,t))C_2(r,t-\tau) - k_d B_2(r,t); \quad r_{l2} \le r \le r_{u2} \qquad (8.1\text{-}3)$$

Eq. (8.1-1) applies over the inner cylinder representing the coating on the stent with the drug, $r_{l1} \le r \le r_{u1}$. Eqs. (8.1-2,3) apply over the outer cylinder representing the arterial wall which receives the drug, $r_{l2} \le r \le r_{u2}$. $C_1(r,t)$ and $C_2(r,t)$ are connected mathematically through BCs at $r = r_{u1} = r_{l2}$.

The PDE dependent and independent variables of eqs. (8.1) are explained in the following Table 8.1.

Variable	Interpretation
$C_1(r,t)$	concentration of the drug in the stent coating
$C_2(r,t)$	concentration of the drug in the arterial wall
$C_2(r,t-\tau)$	delayed $C_2(r,t)$
$B_2(r,t)$	concentration of bound drug in the arterial wall
r	position in the two-cylinder radial domain
t	time

Table 8.1   Dependent, independent variables of DPDEs (8.1)

Eqs. (8.1-1,2) are second order in r and each requires two BCs.

$$\frac{\partial C_1(r = r_{l1}, t)}{\partial r} = 0 \qquad (8.2\text{-}1)$$

$$-D_1\frac{\partial C_1(r = r_{1u})}{\partial r} = -D_2\frac{\partial C_2(r = r_{2l})}{\partial r} \qquad (8.2\text{-}2)$$

$$C_2(r = r_{l2}, t) = K_{cw}C_1(r = r_{u1}, t) \qquad (8.2\text{-}3)$$

$$\frac{\partial C_2(r = r_{u2}, t)}{\partial r} = 0 \qquad (8.2\text{-}4)$$

BCs (8.2) are explained subsequently.

Eqs. (8.1) are first order in t, and each requires one initial condition (IC).

$$C_1(r, t = 0) = f_1(r) \qquad (8.3\text{-}1)$$
$$C_2(r, t = 0) = f_2(r) \qquad (8.3\text{-}2)$$
$$B_2(r, t = 0) = f_3(r) \qquad (8.3\text{-}3)$$

$f_1(r), f_2(r), f_3(r)$ are functions to be specified.

To conclude this discussion of the model, the RHS and LHS terms of eqs. (8.1), (8.2), (8.3) are briefly explained.

The terms in eq. (8.1-1), which define the free drug concentration in the coating, $C_1(r,t)$, are

- $\dfrac{\partial C_1(r,t)}{\partial t}$: Rate of accumulation (term > 0) or depletion (term < 0) of drug in an incremental volume $(2\pi r)(\Delta r)(\Delta z)$ of the coating (z is the axial coordinate in cylindrical coordinates (r,θ,z)).

- $D_1\left(\dfrac{\partial^2 C_1(r,t)}{\partial r^2} + \dfrac{1}{r}\dfrac{\partial C_1(r,t)}{\partial r}\right)$: Net rate of diffusion of the drug into or out of the incremental volume $(2\pi r)(\Delta r)(\Delta z)$ of the coating.

The terms in eq. (8.1-2), which define the free (unbound) drug concentration in the arterial wall, $C_2(r,t)$, are

- $\dfrac{\partial C_2(r,t)}{\partial t}$: Rate of accumulation (term > 0) or depletion (term < 0) of drug in an incremental volume $(2\pi r)(\Delta r)(\Delta z)$ of the arterial wall (again, z is the axial coordinate in cylindrical coordinates (r,θ,z)).

- $D_2\left(\dfrac{\partial^2 C_2(r,t)}{\partial r^2} + \dfrac{1}{r}\dfrac{\partial C_2(r,t)}{\partial r}\right)$: Net rate of diffusion of the drug into or out of the incremental volume $(2\pi r)(\Delta r)(\Delta z)$ of the arterial wall.

- $-k_a(S_0 - B_2(r,t))C_2(r,t-\tau)$: Volumetric rate of transfer of drug in the arterial wall to a bound state. The rate is nonlinear (from the product $B_2(r,t)C_2(r,t-\tau)$) and is based on the delayed concentration $C_2(r,t-\tau)$ to reflect the time required for the binding to occur. S_0 is the initial binding site concentration and therefore $(S_0 - B_2(r,t))$ reflects the decrease in available binding sites with increasing $B_2(r,t)$.

- $+k_d B_2(r,t)$: Volumetric rate of unbinding of the drug which adds to $C_2(r,t)$.

The terms in eq. (8.1-3), which define the bound drug concentration in the arterial wall, $B_2(r,t)$, are

- $\dfrac{\partial B_2(r,t)}{\partial t}$: Rate of accumulation (term > 0) or depletion (term < 0) of bound drug in an incremental volume $(2\pi r)(\Delta r)(\Delta z)$ of the arterial wall (again, z is the axial coordinate in cylindrical coordinates (r,θ,z)).

- $-k_d B_2(r,t)$: Volumetric rate of unbinding of the drug which reduces $B_2(r,t)$.

A diffusion term is not included in eq. (8.1-3) since the bound drug cannot move by diffusion.

Eqs. (8.1-1,2,3) constitute the 3×3 DPDE model. The R routines to implement this model are considered next, starting with the main program.

8.1.1   Main program

The main program is in Listing 8.1.

```
#
# Three PDE model
#
# Delete previous workspaces
  rm(list=ls(all=TRUE))
#
# Access ODE integrator
  library("deSolve");
#
# Access functions for numerical solution
  setwd("f:/dpde/chap8");
  source("pde1a.R");
  source("dss004.R");
  source("dss044.R");
#
# Parameters
  D1=1.0e-02;
  D2=1.0e-02;
  Kcw=1.0e-00;
  tau=1000;
  kd=1.0e-03;
  ka=1.0e+01;
  S0=1.0e-05;
#
# Grids in r
  rl1=140;ru1=190;n1=21;dr1=(ru1-rl1)/(n1-1);
  r1=seq(from=rl1,to=ru1,by=dr1);
  rl2=190;ru2=390;n2=21;dr2=(ru2-rl2)/(n2-1);
  r2=seq(from=rl2,to=ru2,by=dr2);
#
# Initial condition
  u0=rep(0,(n1+2*n2));
  for(i in 1:n1){
    u0[i]=1.0e-05;
  }
  for(i in 1:n2){
    u0[i+n1]   =0;
    u0[i+n1+n2]=0;
  }
#
# Interval in t
  t0=0;tf=400*60*60;nout=26;
  tout=seq(from=t0,to=tf,by=(tf-t0)/(nout-1));
  ncall=0;
#
```

```
# ODE integration
  out=dede(y=u0,times=tout,func=pde1a);
  nrow(out);
  ncol(out);
#
# Store solution
  C1=matrix(0,nrow=n1,ncol=nout);
  C2=matrix(0,nrow=n2,ncol=nout);
  B2=matrix(0,nrow=n2,ncol=nout);
  t=rep(0,nout);
  for(it in 1:nout){
  for(i in 1:n1){
    C1[i,it]=out[it,i+1];
  }
  for(i in 1:n2){
    C2[i,it]=out[it,i+1+n1];
    B2[i,it]=out[it,i+1+n1+n2];
    t[it]=out[it,1];
  }
  }
#
# Display numerical solution
  hrs=60*60;
  iv=seq(from=1,to=nout,by=5);
  for(it in iv){
  cat(sprintf(
    "\n\n        t             r      C1(r,t)"));
  iv1=seq(from=1,to=n1,by=5);
  for(i in iv1){
    cat(sprintf("\n%9.2e%11.2e%12.3e",
      t[it]/hrs,r1[i],C1[i,it]));
   }
  cat(sprintf(
    "\n\n        t             r      C2(r,t)"));
  cat(sprintf(
    "  \n        t             r      B2(r,t)"));
  iv2=seq(from=1,to=n2,by=5);
  for(i in iv2){
    cat(sprintf("\n%9.2e%11.2e%12.3e",
      t[it]/hrs,r2[i],C2[i,it]));
    cat(sprintf("\n%9.2e%11.2e%12.3e\n",
      t[it]/hrs,r2[i],B2[i,it]));
  }
  }
#
# Display ncall
  cat(sprintf("\n\n ncall = %2d",ncall));
#
```

```
# Plot numerical solutions
#
# 2D
  matplot(r1,C1,type="l",xlab="r",ylab="C1(r,t)",
          lty=1,main="",lwd=2,col="black");
  matplot(r2,C2,type="l",xlab="r",ylab="C2(r,t)",
          lty=1,main="",lwd=2,col="black");
  matplot(r2,B2,type="l",xlab="r",ylab="B2(r,t)",
          lty=1,main="",lwd=2,col="black");
#
# 3D
  persp(r1,t,C1,theta=60,phi=45,
        xlim=c(rl1,ru1),ylim=c(t0,tf),xlab="r",ylab="t",
        zlab="C1(r,t)");
  persp(r2,t,C2,theta=45,phi=45,
        xlim=c(rl2,ru2),ylim=c(t0,tf),xlab="r",ylab="t",
        zlab="C2(r,t)");
  persp(r2,t,B2,theta=45,phi=60,
        xlim=c(rl2,ru2),ylim=c(t0,tf),xlab="r",ylab="t",
        zlab="B2(r,t)");
```

Listing 8.1   Main program for eqs. (8.1), (8.2), (8.3).

We can note the following details about Listing 8.1.

- Previous workspaces are deleted.

  ```
  #
  # Three PDE model
  #
  # Delete previous workspaces
    rm(list=ls(all=TRUE))
  ```

- The R ODE integrator library deSolve is accessed. Then the directory with the files for the solution of eqs. (8.1) is designated. Note that setwd (set working directory) uses / rather than the usual \.

  ```
  #
  # Access ODE integrator
    library("deSolve");
  #
  # Access functions for numerical solution
    setwd("f:/dpde/chap8");
    source("pde1a.R");
    source("dss004.R");
    source("dss044.R");
  ```

  pde1a.R is the routine for eqs. (8.1) (discussed subsequently) based on the MOL, a general algorithm for PDEs [1]. dss004, dss044 are library routines for the calculation of first and second spatial derivatives. These routines are listed in Appendix A1 with additional explanation.

- The parameters of eqs. (8.1) are programmed.

```
#
# Parameters
  D1=1.0e-02;
  D2=1.0e-02;
  Kcw=1.0e-00;
  tau=1000;
  kd=1.0e-03;
  ka=1.0e+01;
  S0=1.0e-05;
```

These parameter values are taken in part from Zhu [4, Table 4.1], and have the units of μm for space (1 μm = 1 micron = 1×10^{-6} m) and s for time, for example, $D_1 \Rightarrow \mu m^2/s$. Also, the delay for the DPDEs of eqs. (8.1-2,3), τ, is defined.

- Two radial grids are defined. For the inner cylinder (coating), $r_{l1} = 140 \leq r_1 \leq r_{u1} = 190$, $n_1 = 21$ points with spacing $(190 - 140)/20 = 2.5$ are used so that $r_1 = 140, 142.5, ..., 190$. For the outer cylinder (arterial wall) $r_{l2} = 190 \leq r_2 \leq r_{u2} = 390$, $n_2 = 21$ points are used with spacing $(390 - 190)/20 = 10$ so that $r_2 = 190, 200, ..., 390$.

```
#
# Grids in r
  rl1=140;ru1=190;n1=21;dr1=(ru1-rl1)/(n1-1);
  r1=seq(from=rl1,to=ru1,by=dr1);
  rl2=190;ru2=390;n2=21;dr2=(ru2-rl2)/(n2-1);
  r2=seq(from=rl2,to=ru2,by=dr2);
```

The dimensions are in μm.

- An IC and history vector u0 are defined for n1+2*n2=21+2*21=63 points in r for eqs. (8.3).

```
#
# Initial condition
  u0=rep(0,(n1+2*n2));
  for(i in 1:n1){
    u0[i]=1.0e-05;
  }
  for(i in 1:n2){
    u0[i+n1]   =0;
    u0[i+n1+n2]=0;
  }
```

IC (8.3-1) is constant, 1.0e-05 M (molar). ICs (8.3-2,3) are homogeneous.

- A temporal interval is defined with nout=26 output points in t, and initial and final values of t0=0, tf=400 h or tf = 400*60*60* = 1.44×10^6 s.

```
#
# Interval in t
  t0=0;tf=400*60*60;nout=26;
  tout=seq(from=t0,to=tf,by=(tf-t0)/(nout-1));
  ncall=0;
```

The counter for the calls to pde1a is also initialized.

- The MOL/DODEs for eqs. (8.1) are integrated by the library integrator dede (available in deSolve, [3, Chapter 7]). As expected, the inputs to dede are the IC vector u0, the vector of output values of t, times, and the ODE function, pde1a. The length of u0 (63) informs dede how many DODEs are to be integrated. y, times, func are reserved names.

```
#
# ODE integration
  out=dede(y=u0,times=tout,func=pde1a);
  nrow(out);
  ncol(out);
```

- t is placed in vector t and $C_1(r,t)$, $C_2(r,t)$, $B_2(r,t)$ from eqs. (8.1) are placed in matrices C1, C2, B2 for numerical and graphical display.

```
#
# Store solution
  C1=matrix(0,nrow=n1,ncol=nout);
  C2=matrix(0,nrow=n2,ncol=nout);
  B2=matrix(0,nrow=n2,ncol=nout);
  t=rep(0,nout);
  for(it in 1:nout){
  for(i in 1:n1){
    C1[i,it]=out[it,i+1];
  }
  for(i in 1:n2){
    C2[i,it]=out[it,i+1+n1];
    B2[i,it]=out[it,i+1+n1+n2];
    t[it]=out[it,1];
  }
  }
```

- The solutions of eqs. (8.1) are displayed numerically in r and t with three fors.

```
#
# Display numerical solution
  hrs=60*60;
  iv=seq(from=1,to=nout,by=5);
  for(it in iv){
  cat(sprintf(
    "\n\n        t              r        C1(r,t)"));
```

```
iv1=seq(from=1,to=n1,by=5);
for(i in iv1){
  cat(sprintf("\n%9.2e%11.2e%12.3e",
    t[it]/hrs,r1[i],C1[i,it]));
  }
cat(sprintf(
  "\n\n           t           r       C2(r,t)"));
cat(sprintf(
  "   \n           t           r       B2(r,t)"));
iv2=seq(from=1,to=n2,by=5);
for(i in iv2){
  cat(sprintf("\n%9.2e%11.2e%12.3e",
    t[it]/hrs,r2[i],C2[i,it]));
  cat(sprintf("\n%9.2e%11.2e%12.3e\n",
    t[it]/hrs,r2[i],B2[i,it]));
}
}
```

Every fifth value in t and r are displayed with by=5. Also, t is converted from s to h.

- The counter for the calls to pde1a is displayed at the end of the solution.

```
#
# Display ncall
  cat(sprintf("\n\n ncall = %2d",ncall));
```

- The solutions $C_1(r,t)$, $C_2(r,t)$, $B_2(r,t)$ are plotted in two dimensions (2D) with matplot.

```
#
# Plot numerical solutions
#
# 2D
  matplot(r1,C1,type="l",xlab="r",ylab="C1(r,t)",
          lty=1,main="",lwd=2,col="black");
  matplot(r2,C2,type="l",xlab="r",ylab="C2(r,t)",
          lty=1,main="",lwd=2,col="black");
  matplot(r2,B2,type="l",xlab="r",ylab="B2(r,t)",
          lty=1,main="",lwd=2,col="black");
```

- The solutions $C_1(r,t)$, $C_2(r,t)$, $B_2(r,t)$ are plotted in three dimensions (3D) with persp.

```
#
# 3D
  persp(r1,t,C1,theta=60,phi=45,
        xlim=c(rl1,ru1),ylim=c(t0,tf),xlab="r",ylab="t",
        zlab="C1(r,t)");
  persp(r2,t,C2,theta=45,phi=45,
```

```
        xlim=c(rl2,ru2),ylim=c(t0,tf),xlab="r",ylab="t",
        zlab="C2(r,t)");
    persp(r2,t,B2,theta=45,phi=60,
        xlim=c(rl2,ru2),ylim=c(t0,tf),xlab="r",ylab="t",
        zlab="B2(r,t)");
```

This completes the discussion of the main program in Listing 8.1. The DODE/MOL routine pde1a is considered next.

8.1.2   DODE routine

The DODE/MOL routine pde1a is in Listing 8.2.

```
  pde1a=function(t,u,parm){
#
# Function pde1a computes the t derivative
# vector of C1(r,t), C2(r,t), B2(r,t)
#
# One vector to three vectors
  C1=rep(0,n1);
  C2=rep(0,n2);
  B2=rep(0,n2);
  for(i in 1:n1){
    C1[i]=u[i];
  }
  for (i in 1:n2){
    C2[i]=u[i+n1];
    B2[i]=u[i+n1+n2];
  }
#
# Delayed variable vector
  if (t > tau){
    ulag=lagvalue(t-tau);
  } else {
    ulag=u0;
  }
#
# BC
  C2[1]=Kcw*C1[n1];
#
# C1cr,C2r
  C1r=dss004(rl1,ru1,n1,C1);
  C2r=dss004(rl2,ru2,n2,C2);
#
# BCs
  C1r[1]=0; C1r[n1]=(D2/D1)*C2r[1];
  C2r[n2]=0;
#
# C1rr,C2rr
```

```
  nl=2;nu=2;
  C1rr=dss044(rl1,ru1,nl,C1,C1r,nl,nu);
  nl=1;nu=2
  C2rr=dss044(rl2,ru2,n2,C2,C2r,nl,nu);
#
# C1(r,t-tau),C2(r,t-tau),B2(r,t-tau)
  C1d=rep(0,n1);
  for(i in 1:n1){
    C1d[i]=ulag[i];
  }
  C2d=rep(0,n2);
  B2d=rep(0,n2);
  for(i in 1:n2){
    C2d[i]=ulag[i+n1];
    B2d[i]=ulag[i+n1+n2];
  }
#
# PDEs
  C1t=rep(0,n1);
  C2t=rep(0,n2);
  B2t=rep(0,n2);
  for(i in 1:n1){
    ri=1/r1[i];
    C1t[i]=D1*(C1rr[i]+ri*C1r[i]);
  }
  for(i in 1:n2){
    ri=1/r2[i];
    rate=ka*(S0-B2[i])*C2d[i];
    C2t[i]=D2*(C2rr[i]+ri*C2r[i])-rate+kd*B2[i];
    B2t[i]=rate-kd*B2[i];
  }
#
# Three vectors to one vector
  ut=rep(0,(n1+2*n2));
  for(i in 1:n1){
    ut[i]=C1t[i];
  }
  for(i in 1:n2){
    ut[i+n1]   =C2t[i];
    ut[i+n1+n2]=B2t[i];
  }
#
# Increment calls to pde1a
  ncall<<-ncall+1;
#
# Return derivative vector
  return(list(c(ut)));
}
```

Listing 8.2   DODE/MOL routine for eqs. (8.1), (8.2), (8.3).

We can note the following details about this listing.

- The function is defined.

```
pde1a=function(t,u,parm){
#
# Function pde1a computes the t derivative
# vector of C1(r,t), C2(r,t), B2(r,t)
```

t is the current value of *t* in eqs. (8.1). u is the current numerical solution to eqs. (8.1). parm is an argument to pass parameters to pde1a (unused, but required in the argument list). The arguments must be listed in the order stated to properly interface with dede called in the main program of Listing 8.1. The DODE/MOL approximations of the derivatives $\frac{\partial C_1(r,t)}{\partial t}$, $\frac{\partial C_2(r,t)}{\partial t}$, $\frac{\partial B_2(r,t)}{\partial t}$ of eqs. (8.1) are calculated and returned to dede as explained subsequently.

- The dependent variable vector, u, is placed in three vectors to facilitate the programming of eqs. (8.1).

```
#
# One vector to three vectors
  C1=rep(0,n1);
  C2=rep(0,n2);
  B2=rep(0,n2);
  for(i in 1:n1){
    C1[i]=u[i];
  }
  for (i in 1:n2){
    C2[i]=u[i+n1];
    B2[i]=u[i+n1+n2];
  }
```

- The three dependent variables in u (second input argument of pde1a) are lagged with lagvalue.

```
#
# Delayed variable vector
  if (t > tau){
    ulag=lagvalue(t-tau);
  } else {
    ulag=u0;
  }
```

u is a 63-vector with the three dependent variables of eqs. (8.1) placed according to the ICs programmed in the main program of Listing 8.1. ulag has the lagged values of u. u0, the IC vector, is also the history vector when t <= tau.

- BC (8.2-3) is implemented.

```
#
# BC
  C2[1]=Kcw*C1[n1];
```

K_{cw} is an equilibrium constant that relates the arterial wall concentration $C_2(r = r_{l2}, t) = $ C2[1] to the coating concentration at $r = r_{u1}$, $C_1(r = r_{u1}, t) = $ C1[n1].

- The first derivatives $\dfrac{\partial C_1(r,t)}{\partial r}$, $\dfrac{\partial C_2(r,t)}{\partial r}$ are computed with dss004. The arguments of dss004 are explained in Appendix A1.

```
#
# C1r,C2r
  C1r=dss004(rl1,ru1,n1,C1);
  C2r=dss004(rl2,ru2,n2,C2);
```

- BCs (8.2-1,2,4) are programmed.

```
#
# BCs
  C1r[1]=0; C1r[n1]=(D2/D1)*C2r[1];
  C2r[n2]=0;
```

Subscripts n1, n2 correspond to r_{u1}, r_{u2}, respectively.

- The second derivatives $\dfrac{\partial^2 C_1(r,t)}{\partial r^2}$, $\dfrac{\partial^2 C_2(r,t)}{\partial r^2}$ in eqs. (8.1) are computed with dss044 (listed and discussed in Appendix A1).

```
#
# C1rr,C2rr
  nl=2;nu=2;
  C1rr=dss044(rl1,ru1,n1,C1,C1r,nl,nu);
  nl=1;nu=2
  C2rr=dss044(rl2,ru2,n2,C2,C2r,nl,nu);
```

nl=nu=2 specify Neumann BCs (8.2-1,2) for the coating and nl=1, nu=2 specify Dirichlet and Neumann BCs (8.2-3,4) for the arterial wall for the second derivatives in r.

- The lagged variables $C_1(r, t - \tau)$, $C_2(r, t - \tau)$, $B_2(r, t - \tau)$ are extracted from ulag.

```
#
# C1(r,t-tau),C2(r,t-tau),B2(r,t-tau)
  C1d=rep(0,n1);
  for(i in 1:n1){
    C1d[i]=ulag[i];
```

```
  }
  C2d=rep(0,n2);
  B2d=rep(0,n2);
  for(i in 1:n2){
    C2d[i]=ulag[i+n1];
    B2d[i]=ulag[i+n1+n2];
  }
```

- The MOL programming of eqs. (8.1) steps through the n1=n2=21 values of *r* in two fors.

```
#
# PDEs
  C1t=rep(0,n1);
  C2t=rep(0,n2);
  B2t=rep(0,n2);
  for(i in 1:n1){
    ri=1/r1[i];
    C1t[i]=D1*(C1rr[i]+ri*C1r[i]);
  }
  for(i in 1:n2){
    ri=1/r2[i];
    rate=ka*(S0-B2[i])*C2d[i];
    C2t[i]=D2*(C2rr[i]+ri*C2r[i])-rate+kd*B2[i];
    B2t[i]=rate-kd*B2[i];
  }
```

Branching for $r = 0$ is not required since $r = r_{l1} = 140$ is the smallest value of *r*. The programming of the nonlinear rate in eqs. (8.1-2,3), rate=ka*(S0-B2[i])*C2d[i];, includes the delayed variable $C_2(r, t - \tau)$, that is, $k_a(S_0 = B_2(r, t)C_2(r, t - \tau)$. The correspondence of the PDEs (eqs. (8.1)) and the programming indicates an important feature of the MOL.

- The three derivative vectors C1t, C2t, B2t are placed in one derivative vector, ut, to return to dede (called in the main program of Listing 8.1).

```
#
# Three vectors to one vector
  ut=rep(0,(n1+2*n2));
  for(i in 1:n1){
    ut[i]=C1t[i];
  }
  for(i in 1:n2){
    ut[i+n1]   =C2t[i];
    ut[i+n1+n2]=B2t[i];
  }
```

- The counter for the calls to pde1a is incremented and returned to the main program by «-.

```
#
# Increment calls to pde1a
  ncall<<-ncall+1;
```

- The derivative vector `ut` is returned to `dede` for the next step along the solution.

```
#
# Return derivative vector
  return(list(c(ut)));
}
```

The derivative `ut` is returned as a `list` as required by `dede`. `c` is the R vector utility. The final } concludes `pde1a`.

The concentration $C_1(r,t)$ is reduced in response to BC (8.2-2) as the drug is eluted from the coating, but the coating remains with increasing t so it is termed nondegradable.

The numerical and graphical output from the R routines of Listings 8.1 and 8.2 is considered next.

8.1.3 Numerical, graphical output

```
[1] 26

[1] 64

           t              r         C1(r,t)
    0.00e+00      1.40e+02       1.000e-05
    0.00e+00      1.52e+02       1.000e-05
    0.00e+00      1.65e+02       1.000e-05
    0.00e+00      1.78e+02       1.000e-05
    0.00e+00      1.90e+02       1.000e-05


           t              r         C2(r,t)
           t              r         B2(r,t)
    0.00e+00      1.90e+02       0.000e+00
    0.00e+00      1.90e+02       0.000e+00

    0.00e+00      2.40e+02       0.000e+00
    0.00e+00      2.40e+02       0.000e+00

    0.00e+00      2.90e+02       0.000e+00
    0.00e+00      2.90e+02       0.000e+00

    0.00e+00      3.40e+02       0.000e+00
    0.00e+00      3.40e+02       0.000e+00

    0.00e+00      3.90e+02       0.000e+00
```

```
0.00e+00    3.90e+02    0.000e+00
               .            .
               .            .
               .            .

  Output for t = 80,160,240,320
              removed

               .            .
               .            .
               .            .

        t           r        C1(r,t)
  4.00e+02    1.40e+02    1.540e-06
  4.00e+02    1.52e+02    1.535e-06
  4.00e+02    1.65e+02    1.521e-06
  4.00e+02    1.78e+02    1.500e-06
  4.00e+02    1.90e+02    1.471e-06

        t           r        C2(r,t)
        t           r        B2(r,t)
  4.00e+02    1.90e+02    1.469e-06
  4.00e+02    1.90e+02    1.449e-07

  4.00e+02    2.40e+02    1.301e-06
  4.00e+02    2.40e+02    1.285e-07

  4.00e+02    2.90e+02    1.105e-06
  4.00e+02    2.90e+02    1.093e-07

  4.00e+02    3.40e+02    9.568e-07
  4.00e+02    3.40e+02    9.472e-08

  4.00e+02    3.90e+02    9.036e-07
  4.00e+02    3.90e+02    8.948e-08

ncall = 1953
```

Table 8.2   Abbreviated output from Listings 8.1 and 8.2

We can note the following details about this output.

- 26 output points in t as the first dimension of the solution matrix uout from dede as programmed in the main program of Listing 8.1.

- The solution matrix uout returned by dede has 64 elements as a second dimension. The first element is the value of t. Elements 2–64 in uout are $C_1(r,t)$, $C_2(r,t)$, $B_2(r,t)$ for eqs. (8.1).

- The solution is displayed for $0 \leq t \leq 400$ h, $140 \leq r_1 \leq 190$ μm (for $C_1(r,t)$), $190 \leq r_2 \leq 390$ μm (for $C_2(r,t)$, $B_2(r,t)$).

- Constant IC (8.3-1) and homogeneous (zero) ICs (8.3-2,3) are confirmed (at $t = 0$).

- The computational effort is manageable, `ncall = 1953`, so that `dede` effectively computed a solution to eqs. (8.1).

The details of the solutions are presented in Figures 8.1-1,2,3 as 2D and in Figures 8.1-4,5,6 as 3D.

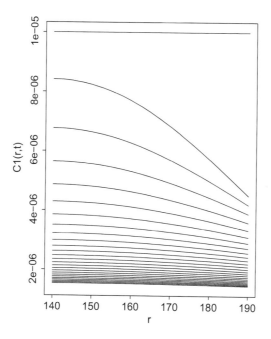

Figure 8.1-1   Numerical solution $C_1(r,t)$ from eq. (8.1-1), `matplot`.

The transition from $C_1(r, t = 0)$=`1.0e-05` to a steady state is clear.
The transition from $C_2(r, t = 0) = 0$ to a steady state is clear.
The transition from $B_2(r, t = 0) = 0$ to a steady state is clear.
The 2D solution of Figure 8.1-1 is confirmed.
The 2D solution of Figure 8.1-2 is confirmed.
The 2D solution of Figure 8.1-3 is confirmed.
Figures 8.1 indicate the complexity of the solutions to eqs. (8.1), (8.2), (8.3).
Numerical experimentation with the model indicated that k_a in eqs. (8.1-2,3) is a sensitive parameter (in determining the magnitude of the nonlinear source term $k_a(S_0 - B_2(r, t))C_2(r, t - \tau)$). As it is increased by factors of 10 (orders of magnitude) in the main program of Listing 8.1, the solution becomes increasingly nonsmooth in r, and eventually unstable in t. These changes are left as an exercise for further study.

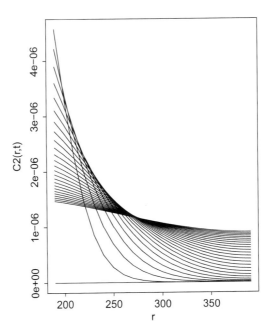

Figure 8.1-2   Numerical solution $C_2(r, t)$ from eq. (8.1-2), `matplot`.

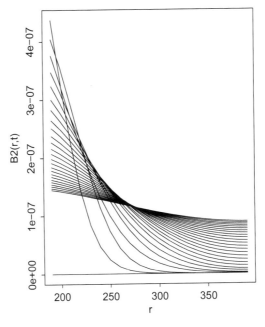

Figure 8.1-3   Numerical solution $B_2(r, t)$ from eq. (8.1-3), `matplot`.

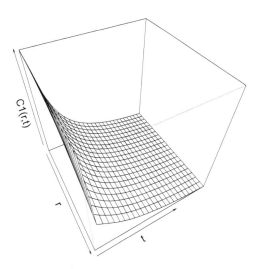

Figure 8.1-4 Numerical solution $C_1(r, t)$ from eq. (8.1-1), persp.

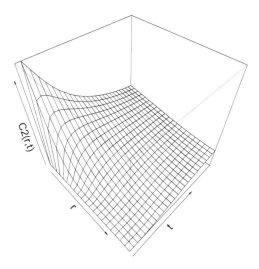

Figure 8.1-5 Numerical solution $C_2(r, t)$ from eq. (8.1-2), persp.

Additional cases for further study are suggested by the preceding results:

- Incremental increase in k_a which defines the contribution (magnitude) of the nonlinear rate $k_a(S_0 - B_2(r, t))C_2(r, t - \tau)$ (rate in pde1a of Listing 8.2). This analysis can include computing and displaying the nonlinear rate

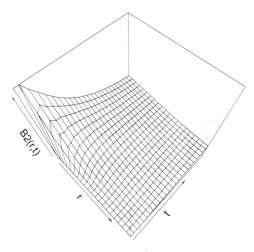

Figure 8.1-6   Numerical solution $B_2(r, t)$ from eq. (8.1-3), `persp`.

(a procedure for computing and displaying individual terms from the DPDEs and BCs is given in [2, Chapter 4]).

- Variation of `tau` to indicate the effect of the delay in eqs. (8.1-2,3) $(C_2(r, t - \tau))$.

- Calculation and display of the flux at the coating–arterial wall interface

$$-D_2 \frac{\partial C_2(r = r_{l2}, t)}{\partial r}.$$

Changes in this flux will affect the arterial wall concentrations, $C_2(r, t)$, $B_2(r, t)$.

This concludes the case of a nondegradable (nonbiodegradable) coating. The next case is for a biodegradable coating in which the diffusivity of the drug varies as described in [4].

8.2   DPDE model, biodegradable stent coating

The calculation of $C_1(r, t), C_2(r, t), B_2(r, t)$ in `pde1a` of Listing 8.2 is based on a constant diffusivity in the coating, D_1. However, for a variable diffusion coefficient to reflect the depletion (erosion) of the coating, D_1 is now variable.

8.2.1 Main program

Eq. (8.1-1) is now

$$\frac{\partial C_1(r,t)}{\partial t} = \frac{1}{r}\frac{\partial \left(rD_1(t)\frac{\partial C_1(r,t)}{\partial r} \right)}{\partial r}$$

$$= D_1(t) \left(\frac{\partial^2 C_1(r,t)}{\partial r^2} + \frac{1}{r}\frac{\partial C_1(r,t)}{\partial r} \right) \qquad (8.1\text{-}4)$$

The only change is $D_1 = D_1(t)$. The equations for $D_1(t)$ are taken from Zhu [4, Table 5.3].

Effective diffusivity:

$$D_1(t) = \frac{(1-\phi)D_s + \kappa\phi D_l}{1-\phi+\kappa\phi} \qquad (8.1\text{-}5)$$

Diffusivity in coating polymer solid:

$$D_s = D_{s0}\left(\frac{M_w}{M_{w0}}\right)^{-\alpha} \qquad (8.1\text{-}6)$$

Coating average molecular weight:

$$M_w = M_{w0}e^{k_w t} \qquad (8.1\text{-}7)$$

Coating porosity change:

$$\phi = \phi_0 + (1-\phi_0)\left(1 + e^{-2kt} - 2e^{-kt}\right) \qquad (8.1\text{-}8)$$

$D_s, D_l, \kappa, D_{s0}, M_{w0}, k_w, k$ are parameters (constants) defined numerically as indicated next.

A minor modification of the programming of eq. (8.1-1) follows to include $D_1(t)$ from eq. (8.1-5), starting with the main program (only the changes in Listing 8.1) are considered.

```
#
# Parameters
  D1=1.0e-02;
  D2=1.0e-02;
  Kcw=1.0e-00;
  tau=1000;
  kd=1.0e-03;
  ka=1.0e+01;
  S0=1.0e-05;
#
```

```
# Variable diffusivity parameters
  Mw0=4.0e+04;
  kw=1.0e-05;
  phi0=0.25;
  k=1.0e-05;
  Ds0=1.0e-05;
  Dl=5.0e-02;
  alpha=1.7;
  kap=1.0e-04;
```

Listing 8.3 Parameters for eqs. (8.1-1,4,5).

These parameters are then passed to `pde1a` and the subordinate routine, `De`, for the effective diffusivity of the drug in the coating. Also, `source("De.R");` is added to the routines that are accessed.

8.2.2 DODE routine

Only the changes in `pde1a` of Listing 8.2 are indicated below.

```
Programming of BCs (8.2-1,2,4)

#
# BCs
  D1v=De(t);
  C1r[1]=0; C1r[n1]=(D2/D1v)*C2r[1];
  C2r[n2]=0;

Programming of eq. (8.1-4)

  for(i in 1:n1){
    ri=1/r1[i];
    D1v=De(t);
    C1t[i]=D1v*(C1rr[i]+ri*C1r[i]);
  }
```

Listing 8.4 Coding for BCs 8.2-1,2,4 and eq. (8.1-4).

The subordinate routine for the effective diffusivity, `De`, is called for BC (8.2-2) and eq. (8.1-4).

8.2.3 Subordinate routine

The subordinate routine `De` for eqs. (8.1-5,6,7,8) follows.

```
  De=function(t){
#
# Function De computes the effective diffusivity
# for the eluted drug in the coating
#
  Mw=Mw0*exp(-kw*t);
  Ds=Ds0*(Mw/Mw0)^{-alpha};
```

```
phi=phi0+(1-phi0)*(1-exp(-2*k*t)-2*exp(-k*t));
De=((1-phi)*Ds+kap*phi*Dl)/(1-phi+kap*phi);
return(c(De));
}
```

Listing 8.5 Routine for eqs. (8.1-5,6,7,8).

The coding follows directly from eqs. (8.1-5,6,7,8). As a test of the revised coding for $D_1(t)$, the previous case for constant D_1 can be included by temporarily placing `De=1.0e-02;` before the `return`. The output is the same as in Table 8.2.

8.2.4 Numerical, graphical output

The abbreviated output from Listings (8.1) to (8.5) follows.

```
[1] 26

[1] 64

         t            r        C1(r,t)
0.00e+00    1.40e+02    1.000e-05
0.00e+00    1.52e+02    1.000e-05
0.00e+00    1.65e+02    1.000e-05
0.00e+00    1.78e+02    1.000e-05
0.00e+00    1.90e+02    1.000e-05

         t            r        C2(r,t)
         t            r        B2(r,t)
0.00e+00    1.90e+02    0.000e+00
0.00e+00    1.90e+02    0.000e+00

0.00e+00    2.40e+02    0.000e+00
0.00e+00    2.40e+02    0.000e+00

0.00e+00    2.90e+02    0.000e+00
0.00e+00    2.90e+02    0.000e+00

0.00e+00    3.40e+02    0.000e+00
0.00e+00    3.40e+02    0.000e+00

0.00e+00    3.90e+02    0.000e+00
0.00e+00    3.90e+02    0.000e+00
                .            .
                .            .
                .            .
  Output for t = 80,160,240,320
            removed
                .            .
```

```
          .                   .
          .                   .
          t                   r          C1(r,t)
     4.00e+02      1.40e+02      1.610e-06
     4.00e+02      1.52e+02      1.610e-06
     4.00e+02      1.65e+02      1.610e-06
     4.00e+02      1.78e+02      1.610e-06
     4.00e+02      1.90e+02      1.610e-06

          t                   r          C2(r,t)
          t                   r          B2(r,t)
     4.00e+02      1.90e+02      1.609e-06
     4.00e+02      1.90e+02      1.583e-07

     4.00e+02      2.40e+02      1.376e-06
     4.00e+02      2.40e+02      1.357e-07

     4.00e+02      2.90e+02      1.097e-06
     4.00e+02      2.90e+02      1.085e-07

     4.00e+02      3.40e+02      8.840e-07
     4.00e+02      3.40e+02      8.765e-08

     4.00e+02      3.90e+02      8.075e-07
     4.00e+02      3.90e+02      8.013e-08

     ncall = 1884
```

Table 8.3   Abbreviated output from Listings 8.1 to 8.5

A comparison of the output in Tables 8.2 and 8.3 indicates that the change from D_1 to $D_1(t)$ substantially changes the solutions to eqs. (8.1) to (8.4). This is confirmed by comparing Figures 8.1 and 8.2.

The transition from $C_1(r, t = 0)$=1.0e-05 to a steady state is clear.

The transition from $C_2(r, t = 0) = 0$ to a steady state is clear.

The transition from $B_2(r, t = 0) = 0$ to a steady state is clear.

The 2D solution of Figure 8.2-1 is confirmed.

The 2D solution of Figure 8.2-2 is confirmed.

The 2D solution of Figure 8.2-3 is confirmed.

Figures 8.1-1,4 and 8.2-1,4 indicate that the change from D_1 to $D_1(t)$ has a substantial effect on $C_1(r, t)$. In particular, at $r = r_{u1} = 190$, the discontinuous change from the IC $C_1(r = r_{u1}, t = 0) = 1.0 \times 10^{-5}$ to BC (8.2-2) with $D_1(t)$ produces a different solution than with D_1 (cf. Figures 8.1-1 and 8.2-1). Also, Figure 8.2-1 has a significant gridding effect (nonsmooth solution) near $r = r_{1u} = 190$. This could be reduced by (1) adding more points to the grid in r (increasing n1) or (2) using a spatial differentiation algorithm that permits a concentration of points near $r = r_{1u} = 190$ such as a spline.

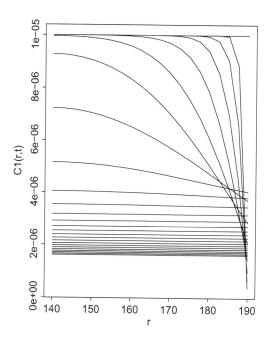

Figure 8.2-1 Numerical solution $C_1(r, t)$ from eqs. (8.1-1,4), `matplot`.

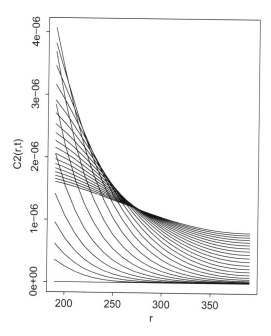

Figure 8.2-2 Numerical solution $C_2(r, t)$ from eq. (8.1-2), `matplot`.

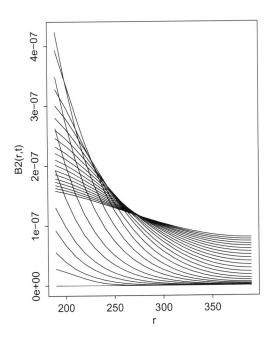

Figure 8.2-3   Numerical solution $B_2(r,t)$ from eq. (8.1-3), `matplot`.

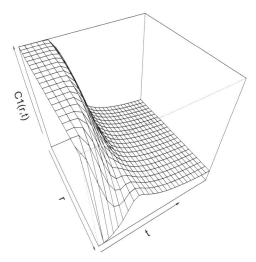

Figure 8.2-4   Numerical solution $C_1(r,t)$ from eqs. (8.1-1,4), `persp`.

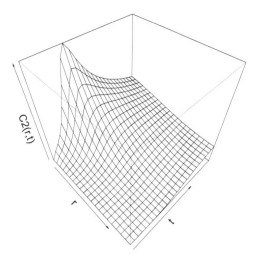

Figure 8.2-5 Numerical solution $C_2(r,t)$ from eq. (8.1-2), `persp`.

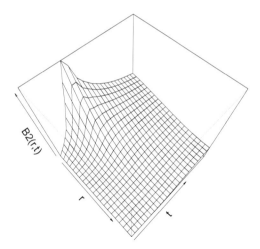

Figure 8.2-6 Numerical solution $B_2(r,t)$ from eq. (8.1-3), `persp`.

The dependent variables $C_1(r,t), C_2(r,t), B_2(r,t)$ appear to be stable and approaching a steady state. For example, $C_1(r,t \to \infty) = 1.610 \times 10^{-6}$ (Table 8.3, Figures 8.2-1,4), $C_2(r,t \to \infty) \approx 1.2 \times 10^{-6}$ (Figures 8.2-2,5), $B_2(r,t \to \infty) \approx 1.2 \times 10^{-7}$ (Figures 8.2-3,6).

8.3   Summary and conclusions

The MOL solution of eqs. (8.1-1,2,3,4) is straightforward as indicated in Listings (8.1)–(8.5). Changes in the coating diffusivity have a significant effect on
the solutions. As a word of caution, changes in the model parameters (constants) might lead to solutions with significant spatial effects (nonsmooth
variations in r), and unstable solutions (failure of the integration in t from
dede). To minimize numerical problems, incremental changes in parameter
values are recommended from the cases for $D_1, D_1(t)$ in Listings 8.1 and 8.3.
   As two possible variants of the two preceding cases, we could consider

- A coating diffusivity that is a function of $C_1(r,t)$, that is, $D_1(C_1(r,t))$. This
  leads to a nonlinear form of eq. (8.1-1)

$$\frac{\partial C_1(r,t)}{\partial t} = \frac{1}{r}\frac{\partial\left(rD_1(C_1(r,t))\dfrac{\partial C_1(r,t)}{\partial r}\right)}{\partial r} \tag{8.1-9}$$

or with expansion of the nonlinear diffusion term

$$\frac{\partial C_1(r,t)}{\partial t} = D_1(C_1(r,t))\frac{\partial^2 C_1(r,t)}{\partial r^2}$$
$$+ \frac{dD_1(C_1(r,t))}{dC_1(r,t)}\frac{\partial C_1(r,t)}{\partial r} + \frac{1}{r}D_1(C_1(r,t))\frac{\partial C_1(r,t)}{\partial r} \tag{8.1-10}$$

  A requirement for using eq. (8.1-10) is the programming of the derivative
  $\dfrac{dD_1(C_1(r,t))}{dC_1(r,t)}$ in the MOL/DODE routine. For example, if $D_1(C_1(r,t)) =$
  $a_0 + a_1 C_1(r,t)$ (a linear function of $C_1(r,t)$), then $\dfrac{dD_1(C_1(r,t))}{dC_1(r,t)} = a_1$.
  Otherwise, the programming of eq. (8.1-10) is a straightforward exten-
  sion of the programming of eq. (8.1-1) in Listing 8.2.

- Eqs. (8.1) to (8.4) are 1D in r. They can be extended dimensionally by
  adding the other two components of cylindrical coordinates, θ and z. θ
  could be used to account for the angular variation of the drug concentration, possibly resulting from the struts of the stent [4, p. 41]. z could be
  used to account for the axial distance along the stent. This could include
  convection of the drug if a PDE for the bloodstream is added to the
  model.

Eq. (8.1-1) extended to 3D is

$$\frac{\partial C_1(r,\theta,z,t)}{\partial t} = D_{1r}\left(\frac{\partial^2 C_1(r,\theta,z,t)}{\partial r^2} + \frac{1}{r}\frac{\partial C_1(r,\theta,z,t)}{\partial r}\right)$$

$$+ D_{1\theta}\frac{1}{r^2}\frac{\partial^2 C_1(r,\theta,z,t)}{\partial \theta^2} + D_{1z}\frac{\partial^2 C_1(r,\theta,z,t)}{\partial z^2};$$

$$r_{l1} \leq r \leq r_{u1}; \ \theta_{l1} \leq \theta \leq \theta_{u1}; z_{l1} \leq z \leq z_{u1} \qquad (8.1\text{-}11)$$

Eq. (8.1-11) is second order in θ and z, and requires two BCs for each dimension.

Eqs. (8.1-2,3) will also include derivatives in θ and z and associated BCs. An example of a MOL solution of a 2D PDE system is given in [1, p. 148]. The extension of the 1D model to 2D and 3D is left as an exercise.

References

[1] Schiesser, W.E. (2016), *Method of Lines PDE Analysis in Biomedical Science and Engineering*, Wiley, Hoboken, NJ.

[2] Schiesser, W.E. (2019), *Numerical PDE Analysis of Retinal Neovascularization*, Elsevier, Cambridge, MA.

[3] Soetaert, K., J. Cash, and F. Mazzia (2012), *Solving Differential Equations in R*, Springer-Verlag, Heidelberg, Germany.

[4] Zhu, X. (2015), *Mathematical Modeling and Simulation of Intravascular Drug Delivery from Drug-Eluting Stents with Biodegradable PLGA Coating*, PhD Thesis, Department of Chemical Engineering, MIT, Cambridge, MA.

Appendix A1: Functions **dss004, dss044**

A1.1 Function **dss004**

A listing of function dss004 follows.

```
  dss004=function(xl,xu,n,u) {
#
# An extensive set of documentation comments detailing
# the derivation of the following fourth order finite
# differences (FDs) is not given here to conserve
# space.  The derivation is detailed in Schiesser,
# W. E., The Numerical Method of Lines Integration
# of Partial Differential Equations, Academic Press,
# San Diego, 1991.
#
# Preallocate arrays
  ux=rep(0,n);
#
# Grid spacing
  dx=(xu-xl)/(n-1);
#
# 1/(12*dx) for subsequent use
  r12dx=1/(12*dx);
#
# ux vector
#
# Boundaries (x=xl,x=xu)
  ux[1]=r12dx*(-25*u[1]+48*u[  2]-36*u[  3]+16*u[  4]-
3*u[  5]);
  ux[n]=r12dx*( 25*u[n]-48*u[n-1]+36*u[n-2]-16*u[n-3]+
3*u[n-4]);
#
# dx in from boundaries (x=xl+dx,x=xu-dx)
  ux[  2]=r12dx*(-3*u[1]-10*u[  2]+18*u[  3]-6*u[  4]+u[  5]);
  ux[n-1]=r12dx*( 3*u[n]+10*u[n-1]-18*u[n-2]+6*u[n-3]-u[n-4]);
#
# Interior points (x=xl+2*dx,...,x=xu-2*dx)
  for(i in 3:(n-2))ux[i]=r12dx*(-u[i+2]+8*u[i+1]-8*u[i-1]+
u[i-2]);
#
```

```
# All points concluded (x=xl,...,x=xu)
  return(c(ux));
}
```

The input arguments are

> xl   lower boundary value of x

> xu   upper boundary value of x

> n   number of points in the grid in x,
>   including the end points

> u   dependent variable to be differentiated,
>   an n-vector

The output, ux, is an n-vector of numerical values of the first derivative of u.

The finite difference (FD) approximations are a weighted sum of the dependent variable values. For example, at point i

```
for(i in 3:(n-2))ux[i]=r12dx*(-u[i+2]+8*u[i+1]-8*u[i-1]+
u[i-2]);
```

The weighting coefficients are -1, 8, 0, -8, 1 at points i-2, i-1, i, i+1, i+2, respectievly. These weighting coefficients are antisymmetric (opposite sign) around the center point i because the computed first derivative is of odd order. If the derivative is of even order, the weighting coefficients would be symmetric (same sign) around the center point (as in dss044 that follows).

For i=1, the dependent variable at points i=1,2,3,4,5 is used in the FD approximation for ux[1] to remain within the x domain (fictitious points outside the x domain are not used).

```
ux[1]=r12dx*(-25*u[1]+48*u[2]-36*u[3]+16*u[4]-3*u[5]);
```

Similarly, for i=2, points i=1,2,3,4,5 are used in the FD approximation for ux[2] to remain within the x domain (fictitious points outside the x domain are avoided).

```
ux[2]=r12dx*(-3*u[1]-10*u[2]+18*u[3]-6*u[4]+u[5]);
```

At the right boundary $x = x_u$, points at i=n,n-1,n-2,n-3,n-4 are used for ux[n],ux[n-1] to avoid points outside the x domain.

In all cases, the FD approximations are fourth order correct in x.

A1.2   Function dss044

A listing of function dss044 follows.

```
  dss044=function(xl,xu,n,u,ux,nl,nu) {
#
# The derivation of the finite difference
# approximations for a second derivative are
# in Schiesser, W. E., The Numerical Method
# of Lines Integration of Partial Differential
# Equations, Academic Press, San Diego, 1991.
#
# Preallocate arrays
  uxx=rep(0,n);
#
# Grid spacing
  dx=(xu-xl)/(n-1);
#
# 1/(12*dx**2) for subsequent use
  r12dxs=1/(12*dx^2);
#
# uxx vector
#
# Boundaries (x=xl,x=xu)
  if(nl==1)
    uxx[1]=r12dxs*
           (45*u[  1]-154*u[  2]+214*u[  3]-
           156*u[  4] +61*u[  5] -10*u[  6]);
  if(nu==1)
    uxx[n]=r12dxs*
           (45*u[  n]-154*u[n-1]+214*u[n-2]-
           156*u[n-3] +61*u[n-4] -10*u[n-5]);
  if(nl==2)
    uxx[1]=r12dxs*
           (-415/6*u[  1] +96*u[  2]-36*u[  3]+
             32/3*u[  4]-3/2*u[  5]-50*ux[1]*dx);
  if(nu==2)
    uxx[n]=r12dxs*
           (-415/6*u[  n] +96*u[n-1]-36*u[n-2]+
             32/3*u[n-3]-3/2*u[n-4]+50*ux[n]*dx);
#
# dx in from boundaries (x=xl+dx,x=xu-dx)
    uxx[  2]=r12dxs*
           (10*u[  1]-15*u[  2]-4*u[  3]+
            14*u[  4]- 6*u[  5]  +u[  6]);
    uxx[n-1]=r12dxs*
           (10*u[  n]-15*u[n-1]-4*u[n-2]+
            14*u[n-3]- 6*u[n-4]  +u[n-5]);
#
# Remaining interior points (x=xl+2*dx,...,
# x=xu-2*dx)
  for(i in 3:(n-2))
    uxx[i]=r12dxs*
```

```
            (-u[i-2]+16*u[i-1]-30*u[i]+
        16*u[i+1]    -u[i+2]);
#
# All points concluded (x=xl,...,x=xu)
  return(c(uxx));
}
```

The input arguments are

> xl   lower boundary value of x

> xu   upper boundary value of x

> n   number of points in the grid in x, including the end points

> u   dependent variable to be differentiated, an n-vector

> ux   first derivative of u with boundary condition (BC) values, an n-vector

> nl   type of BC at x=xl
> 1: Dirichlet BC
> 2: Neumann BC

> nu   type of BC at x=xu
> 1: Dirichlet BC
> 2: Neumann BC

The output, uxx, is an n-vector of numerical values of the second derivative of u.

The FD approximations are a weighted sum of the dependent variable values. For example, at point i

```
for(i in 3:(n-2))
  uxx[i]=r12dxs*
        (-u[i-2]+16*u[i-1]-30*u[i]+
        16*u[i+1]    -u[i+2]);
```

The weighting coefficients are -1, 16, -30, 16, -1 at points i-2, i-1, i, i+1, i+2, respectievly. These weighting coefficients are symmetric around the center point i because the computed second derivative is of even order. If the derivative is of odd order, the weighting coefficients will be antisymmetric (opposite sign) around the center point.

For nl=2 and/or nu=2 the boundary values of the first derivative are included in the FD approximation for the second derivative, uxx. For example, at x=xl (with nl=2),

```
if(nl==2)
  uxx[1]=r12dxs*
            (-415/6*u[  1] +96*u[  2]-36*u[  3]+
             32/3*u[  4]-3/2*u[  5]-50*ux[1]*dx);
```

In computing the second derivative at the left boundary, uxx[1], the first derivative at the left boundary is included, that is, ux[1]. In this way, a Neumann BC is accommodated (ux[1] is included in the input argument ux).

For nl=1, only values of the dependent variable (and not the first derivative) are included in the weighted sum.

```
if(nl==1)
  uxx[1]=r12dxs*
            (45*u[  1]-154*u[  2]+214*u[  3]-
             156*u[  4] +61*u[  5] -10*u[  6]);
```

The dependent variable at points i=1,2,3,4,5,6 is used in the FD approximation for uxx[1] to remain within the x domain (fictitious points outside the x domain are not used).

Six points are used rather than five (as in the centered approximation for uxx[i]) since the FD applies at the left boundary and is not centered (around i). Six points provide a fourth-order FD approximation which is the same order as the FDs at the interior points in x.

Similar considerations apply at the upper boundary value of x with nu=1,2.

Robin boundary conditions (BCs) can also be accommodated with nl=2, nu=2. In all three cases, Dirichlet, Neumann, and Robin, the BCs can be linear and/or nonlinear.

Additional details concerning dss004, dss044 are available from Griffiths and Schiesser [1].

Reference

[1] Griffiths, G.W., and W.E. Schiesser (2012), *Traveling Wave Analysis of Partial Differential Equations*, Elsevier/Academic Press, Boston, MA.

Index

Printed and bound by CPI Group (UK) Ltd, Croydon, CR0 4YY

17/10/2024

01775682-0011